# THE LOGIC OF
# OUR LANGUAGE

# THE LOGIC OF OUR LANGUAGE

## An Introduction to Symbolic Logic

Rodger L. Jackson
and Melanie L. McLeod

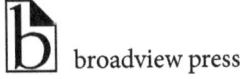

© 2015 Rodger L. Jackson and Melanie L. McLeod

All rights reserved. The use of any part of this publication reproduced, transmitted in any form or by any means, electronic, mechanical, photocopying, recording, or otherwise, or stored in a retrieval system, without prior written consent of the publisher—or in the case of photocopying, a licence from Access Copyright (Canadian Copyright Licensing Agency), One Yonge Street, Suite 1900, Toronto, Ontario M5E 1E5—is an infringement of the copyright law.

LIBRARY AND ARCHIVES CANADA CATALOGUING IN PUBLICATION

Jackson, Rodger L., author
    The logic of our language : an introduction to symbolic logic /
Rodger L. Jackson and Melanie L. McLeod.

Includes index.
ISBN 978-1-55481-184-7 (pbk.)

    1. Logic, Symbolic and mathematical—Textbooks. 2. Logic, Symbolic and mathematical—Problems, exercises, etc. I. McLeod, Melanie L., author II. Title.

BC135.J32 2014                160            C2014-905251-0

BROADVIEW PRESS is an independent, international publishing house, incorporated in 1985.

We welcome comments and suggestions regarding any aspect of our publications—please feel free to contact us at the addresses below or at broadview@broadviewpress.com.

NORTH AMERICA
Post Office Box 1243
Peterborough, Ontario
K9J 7H5  Canada

555 Riverwalk Parkway
Tonawanda, NY 14150, USA
tel: (705) 743–8990
fax: (705) 743–8353

customerservice@broadviewpress.com

UK, EUROPE, CENTRAL ASIA, MIDDLE EAST, AFRICA, INDIA, AND SOUTHEAST ASIA
Eurospan Group, 3 Henrietta St., London WC2E 8LU, United Kingdom
tel: 44 (0) 1767 604972  fax: 44 (0) 1767 601640
eurospan@turpin-distribution.com

AUSTRALIA AND NEW ZEALAND
NewSouth Books
c/o TL Distribution, 15–23 Helles Ave.
Moorebank, NSW 2170, Australia
tel: (02) 8778 9999  fax: (02) 8778 9944
orders@tldistribution.com.au

www.broadviewpress.com

Broadview Press acknowledges the financial support of the Government of Canada through the Canada Book Fund for our publishing activities.

Edited by Robert M. Martin
Typesetting by Em Dash Design

Printed in Canada

# Contents

Instructor's Preface .................................................................................................7

## UNIT ONE: WHAT DOES IT MEAN?
### Patterns of Statements
Chapter One: Logic and Languages .......................................................................11
Chapter Two: Names and Predicates ....................................................................19
Chapter Three: Quantifiers ...................................................................................33
Chapter Four: Negations and Conjunctions ........................................................51
Chapter Five: Conditionals and Disjunctions .....................................................69
Chapter Six: Biconditionals and Identity ............................................................87
Unit One Review ...................................................................................................105
Unit One: Answers to Selected Problems ..........................................................109

## UNIT TWO: IS IT TRUE?
### Properties and Relations of Statements
Chapter Seven: Connectives and Truth Tables ..................................................121
Chapter Eight: Truth Trees ..................................................................................153
Chapter Nine: Relationships between Statements............................................173
Chapter Ten: Reintroducing Names, Predicates, Quantifiers, and Identity ........187
Unit Two Review ...................................................................................................203
Unit Two: Answers to Selected Problems...........................................................209

## UNIT THREE: IS IT VALID?
### Patterns and Properties of Arguments
Chapter Eleven: Arguments, Trees, and Tables .................................................245
Chapter Twelve: Method of Proof ........................................................................267
Chapter Thirteen: Proof Rules for Quantifiers ..................................................291
Unit Three Review .................................................................................................307
Unit Three: Answers to Selected Problems ........................................................311

Postscript: Logic in Real Life ...............................................................................333

Index ........................................................................................................................345

# Instructor's Preface

Over many years of teaching logic, we've noticed a few things. The first is that introductory logic classes are not overpopulated with aspiring logicians or philosophy students. Some students take the course because they have a quantitative reasoning requirement, and they don't want to take any math. Or because the course would count as a "breadth" or "at some distance" or "distribution" requirement. On occasion, math or science majors may use it as a cognate. Often it is taken because it meets the very important "fits into my schedule" requirement.

This being the case, our goal was to develop a course for the general public that makes up the student body: an accessible introduction to the intricacies of symbolic logic. We first thought about the way we learned logic, and the way most logic texts begin: with what are called "basic concepts," such as 'argument' and 'validity.' We realized that while these are *fundamental* logical concepts, they were hardly *elementary*, Watson ("So, in valid arguments the statements are always true?" "So, in a valid argument the conclusion is false when the premises are false?" "So, premises always have to be valid?").

Another thing we both remembered from our early days was slogging through some very austere prose. Explanations were not always as clear as they could be. Moreover, while historical and theoretical background may be interesting to some philosophy students, it could get in the way of mastering the basic skills of logic. Finally, the books we learned from and those we've taught with over the years, had students trying to master the most complicated translations at the same time as they were trying to master the more difficult proofs in the logic of quantifiers and identity. And this usually in the last couple weeks of term.

So, what with our varied potential audience and with these impediments to the mastery of introductory symbolic logic, we decided to turn our logic course on its head.

We begin the book, not with what is basic to philosophers, but with what is basic and common to all our students—they speak a language and know the basics of that language. They know the difference between a subject and a predicate—or they do with a couple minutes of review. They know what an 'and' statement is, and a lot more. They also know that as college students, they often have to read and interpret language of a greater complexity that they're used to, no matter what their field of study. What they don't know, and what symbolic logic can show them, is the logical structure of their language. Unit 1, then, deals with translation between English and the symbolic language, starting with grammatically simple statements and ending with compound statements. From their solid foundation as language users, the students learn that the grammatically "simple" can be very complex indeed. Then throughout the book, they continue practicing translation, while adding other logical skills.

In Unit 2, we introduce a more simplified logical language that lets them focus on the truth-functional nature of connectives. We use both truth tables and trees to explore the property of truth and the truth-relations that hold between statements, ending with implication and equivalence. In this way, when they move on to Unit 3 and argument evaluation, they've already worked with implication, and they can *see* that the property of validity depends on the logical relation of implication. In Unit 3, we use short-form tables and trees to determine validity, before moving on to the method of proof.

Finally, in a short post-script, we show some practical application of those skills they've worked on throughout the term. We introduce an Argument-Analysis tree, for evaluating longer English language arguments, then go through a detailed example of argument evaluation. We also provide five arguments, ranging in length from one to several paragraphs, so they can practice with argument analysis.

Our approach is to emphasize skills throughout the course, and have students practice them constantly throughout the term. The reason for this is very simple— we want them to understand that there are recurring patterns in language and in reasoning and that by seeing and working with these patterns in a logic course, one will be more likely to see them, understand them, and deal with them appropriately when they're encountered in other courses or outside the classroom.

But we also want this text to be flexible. As our approach is not "theory heavy," there is plenty of opportunity for the instructor to introduce topics in philosophy of logic that are of interest to them or to their students. The structure we have chosen for dealing with tables, trees, and proofs goes from "machine thinking," to the quasi-mechanical, to the creative skill of natural deduction; and, in the post-script, to the even more creative job of reconstructing and then evaluating longer arguments. But this sequence need not be adhered to. For example, short-form tables may be introduced earlier, e.g., when introducing implication and equivalence, or not be used at all. Or after introducing the concept of the argument, the instructor could move immediately to proofs. There are plenty of practice problems in the book to keep most students challenged throughout the term, whether in class or as homework. In the materials available to instructors, you have resources for many more problem sets in areas you might want to emphasize.

We enjoy teaching logic and watching our students gain the valuable skills that a study of formal logic provides. We are excited to have this opportunity to share our approach with other logic teachers and hope you find it as effective in your class as we have.

# UNIT ONE: WHAT DOES IT MEAN?

# Patterns of Statements

# Chapter One: Logic and Languages

> **What's Up?**
> Logic and Symbolic Logic
> Natural and Artificial Language
> Sentences and Statements

## WHAT IS LOGIC?

Often students who find themselves in a logic class don't have any clear idea what logic is. But probably all those students have used the words 'logic,' 'logical,' and 'illogical,' and used them correctly. For example, you might describe your Aunt Abigail as illogical if she refused to keep her money in the bank because, she reasons, that's where the bank robbers go. Or you might describe your Uncle Ulysses as being logical when he gets all his materials and blueprints together before he starts building his birdhouses.

In general, you describe someone as logical if what they say or do makes sense. You call their decisions or opinions logical if they have good reasons for those decisions or opinions. You call someone illogical if their decisions, actions, or beliefs seem senseless or have no rhyme or reason to them. You might also correctly apply the words 'logical' or 'illogical' to describe books, movies, government policies, or op-ed pieces in the paper.

This everyday use of the term 'logic' gives us a good start for a more formal definition of logic. To start, we can look at logic from the point of view of a thinker. Logic is both a study and a discipline about rational thinking. Is our thinking reasonable? Do our decisions make sense? Are our explanations satisfactory? And importantly, what *criteria* do we use to answer these questions?

Logic can also be about the *character* of the thinker. In this way it is a study and discipline about becoming a critical thinker. Is the thinker engaged—does he or she question beliefs and assumptions? Is the thinker open to new information and constructive criticism? Does the thinker actively try to self-correct his or her thinking by examining his or her personal or cultural biases? Does the thinker "re-think"—trying to recast an issue using different concepts or categories?

The second way of looking at logic is from the viewpoint of a community of thinkers. In this way, logic is a study and discipline about how we *communicate* thinking. Its focus is on the public and communal *language* that *expresses* thinking. What do our different words and sentences express? How do they express them?

What is a correct response to certain sentences? Is a sentence true or not? When sets of sentences are combined in a discourse, are they *reasonably* combined? Do they express coherent thought? Are all the sentences true? Logic can also look at the *manner* in which thought is expressed. Is the language clear, precise, and unambiguous? Is the language appropriate to the "target audience"? Is the language fair—or does it misrepresent what it's talking about?

In practice, of course, these two sides of logic come together. We want to be able to think clearly and reasonably so we can make good decisions, reach our goals, and defend ourselves against people or circumstances that may work to lead us astray. But we don't just live in our heads; thankfully, the world is full of fellow thinkers. We may have great ideas, but we need to be able to communicate them in understandable ways to friends, teachers, bosses, and our fellow citizens of the world. Our thinking can be honed and improved by the input of others who communicate with us, and we can benefit others by a serious consideration, followed by a reasoned response, to their views.

The study, practice, and discipline of logic help with all of this. On those rare occasions when people think about what logicians or philosophers do, they often think of an esoteric intellectual practice, shared perhaps with other esoteric intellectuals, but not of much use in the everyday world. And, to be fair, this can be the case. But we all think, and we are surrounded by other thinkers. We have to make decisions that will have significant impact on our lives and the lives of others. We are also subject to the decisions of others, and we need to be able to evaluate public policies and proposals—and be able to critically and publicly respond to them. And, of course, in our college careers, we are often asked to learn and evaluate material of much greater complexity than we've tackled before. Logic can help us understand and sort out this material. Logic, then, is an extremely practical discipline that touches every aspect of our thinking lives.

So, what is logic? As we've seen, logic encompasses a great deal: logic is a study, a practice, and a discipline; it's about thought and language; it's about an individual and a community. On this view, logic is pretty wide ranging, and has many aspects. While it's risky to give short, simple definitions for what is complex, such definitions are useful, and we offer this one: Logic is the study of structures and patterns that express meaning and reasoning.

## WHAT IS SYMBOLIC LOGIC?

In this course in symbolic logic, our focus is on certain structures of language: sentences and groups of sentences. In a *natural language* (such as English, Spanish, Chinese, Hindi, or Tagalog, to name just a few), words can have multiple meanings, sentences can be stated in different ways, and

> **Symbolic Logic**
> Study of structures and their patterns that express statements and reasoning.

groups of sentences can be arranged in any order. In symbolic logic we learn a *simplified artificial language* that we use to show the *logical* structures and patterns of a natural language. This language of symbols strips away all the nuance and beauty (or, on the other hand, all the ambiguity and confusion) of our spoken and written language. It also strips away content: thus the symbolic sentence A & B can represent the natural language sentences "It's raining and it's Tuesday," or "Bill is happy but that won't last." These features of symbolic logic language are useful for displaying the logical form of natural language sentences.

While there are many different types of sentences in a natural language, our focus will be on *declarative sentences*—those structures of the language that express claims or statements. And, while there are many meaningful groups of sentences (in, for example, novels, poems, essays, reports, textbooks), we focus on a particular structure that expresses reasoning: the *deductive argument*.

## WHY STUDY SYMBOLIC LOGIC?

This may strike you as a very narrow focus, especially after learning of the scope of logic. And you'd be right. It also might strike you that rather than having to learn and use a quite different artificial language, we should be able to do logic in our own language—that that would be useful—and again you'd be right. Also true is the fact that other kinds of sentences (such as questions or commands) warrant consideration. Often in informal logic or critical thinking courses, study includes these different kinds of sentences, as well as inductive, abductive, and convergent arguments, and they use a natural language to do the job. We heartily recommend such courses, and indeed believe that a good foundation in logic should include a critical thinking course.

But there are good reasons for complementing such courses with a foundation in symbolic logic. The first is that the level of analysis in symbolic logic is usually deeper. This lets us concentrate on the logical form of statements, allows us to determine the truth conditions for certain kinds of statements, and lets us determine if our reasoning "hangs together" in a logical way.

We speak our language every day, and we think in it too. But it is so familiar and automatic that often we don't think *about* it. And, often, this works well enough. But think about those times, for instance, when you've had to tackle complex or unfamiliar material. At those times, you recognize what you're reading as English, you may know and recognize many of the words, but you're still not sure what it means, how it hangs together, where it's going, or what's the point. The skills you master in symbolic logic can help you gain confidence in approaching new and complex material: with the experience of recognizing the patterns and flow of discourse, and the knowledge of methods for analyzing and representing discourse, while you still might have to struggle with the content of some particular thing you hear or read, the structure of new material will be far less intimidating. In short, symbolic logic gives you the skills to recast complex material to make it more meaningful to you, which makes it easier to learn and evaluate.

Another good reason for studying symbolic logic is that it gives you skill in manipulating an artificial language, which can serve as an excellent foundation for studying other symbolic systems, such as computer languages or the languages of math or science. Understanding that all languages have a logical structure can also help if you need to acquire a reading knowledge of a second natural language.

## LANGUAGE: NATURAL AND ARTIFICIAL

We can begin our study of logic with something you already know: the English language. Think of it, with only 26 letters we can form a million words or more. We also have rules of grammar that allow us to create countless sentences on any imaginable subject.

A natural language, English and all others, develops with time and use. It is used in the whole range of human encounters: personal relationships, business transactions, politics, arts, sciences, etc. Some terms and sentences in a natural language may be understood only by select groups (e.g., the technical language of astrophysics or the slang used by teenagers), and some terms and sentences are so simple that two-year-olds understand and use them. Natural languages allow us to express ourselves with style; we can be dull and to the point, or elaborate and circumlocutory, or even dull and circumlocutory. Natural languages are adaptable and can be beautiful. But, they also have problems, such as vagueness, ambiguity, and imprecision.

> Artificial language for symbolic logic uses a simplified vocabulary and grammar to unambiguously demonstrate the logical structure, properties, and relationships of statements.

An artificial language, unlike a natural language, is *constructed for* a particular purpose, rather than "organically grown" through use in a language community. In the case of the artificial language we use in this course, it is constructed for the purpose of *unambiguously representing the logical structure of natural language statements*. When we've shown this logical structure, we can then evaluate important properties of statements and arguments, as well as relationships between or among statements. Like any language, our artificial language has a vocabulary and a grammar, but these are much simpler than those of many natural languages (you'll find no irregular verbs in our language!).

## SENTENCES AND STATEMENTS

We use sentences in both natural and artificial language. Let's take a moment to review some English language sentences. The table below shows common sentence types and an example of each type.

## Table 1A

| Sentence Type | Example |
| --- | --- |
| a) Declarative | a) Everybody loves somebody. |
| b) Interrogative | b) Why do you like coffee so much? |
| c) Imperative | c) Shut the door. |
| d) Exclamatory | d) Dude! |

Declarative sentences make statements. Interrogatives express questions, imperatives express commands, and exclamations "cry out."

It is important to be able to differentiate among these types of sentences, and to understand the roles they play in our language. Moreover, we distinguish kinds of declarative sentences: ones that express opinions, definitions, widely accepted facts, assumptions, well-known falsehoods, and so on. In a critical thinking course, students learn to analyze how these various sentence types function in reasoning.

In this course in symbolic logic, we are concerned only with declarative sentences. More explicitly, we are concerned with the *statement or proposition that the declarative sentence expresses*. We define a statement as *an assertion that has a truth value*. In other words, a statement is either true or it is false. We will learn to construct declarative sentences in our artificial language; that is, we will express statements using the artificial language.

Although the declarative sentence and the statement are related, there are important differences between the two concepts. In order to see this, let's take a simple example of an English language declarative sentence, "Our oak tree is taller than our neighbor's elm." We can consider this sentence from two distinct perspectives: the *content* of the sentence and the *particular way* in which the content is being expressed. To see the difference between these two consider the following five alternate ways of expressing the same content as our sentence about the oak:

1. The oak tree we own is taller than the elm tree owned by our neighbor.
2. Our neighbor's elm is shorter than our oak.
3. Our neighbor's elm isn't as tall as our oak tree.
4. Our oak's taller than the neighbor's elm.
5. The neighbor's elm tree's shorter than our oak.

While all of these sentences are saying the same thing, that is, expressing the same statement, they are doing it in different ways. We are not even restricted to these particular letters or punctuation or words, of course, since we could just as easily have used Arabic or Thai or any other natural language to express the same content. The underlying meaning or *propositional content* of the declarative sentence remains the same in each case, and the content is what we call a statement. Hence, one distinction between statements and declarative sentences is that *multiple declarative sentences can express the same statement*.

As a matter of fact, in English we sometimes even use sentences *other than* declaratives to express statements. For instance, your neighbor might say to you: "Look at how my elm is shorter than your oak," and use the imperative form. Or you might say to your neighbor: "Don't you wish *your* elm was a tall as *my* oak?" and use the interrogative form. In *context*, for native speakers, we can usually sort out these "statements in disguise," and correctly interpret the use of the imperative or interrogative form as a stylistic or rhetorical flourish (or occasionally, as indicative of the temperament of the speaker!).

Further, without going into any of the many challenging philosophical issues surrounding the process of translation, at the very least *a good translation of a declarative sentence from one language to another ought to unambiguously capture the statement*. If you translated the above sentence and didn't capture the idea that the oak is taller than the elm, then you'd have misrepresented the meaning. You'd need to fix your translation to make sure it successfully expressed the original statement. This means there can be better and worse translations.

Keep these points in mind throughout the text: there can be more than one way of expressing the same proposition, and there are more or less successful translations of the proposition from one language to another. Good translation from one language to another is a skill that can be learned and an art that develops as you gain facility with both languages.

An important first step is to strive for fluency in both languages; and you're in luck here as you are probably already fluent in English if you're reading this text. This text provides many examples of translations between the English language and our constructed artificial language. Do study them carefully; be sure you can always tell why we have translated a statement as we do. Do as many of the practice exercises as you can: the more practice, the more you'll recognize language patterns and the quicker and more accurately you will translate. And if you run out of problems to do in this text, look at texts from your other classes—try using your translation skills on passages you're reading for those classes. This will not only give you more logic practice, you may find that you are understanding those other readings better because you've slowed down and are really trying to find out what they mean.

Our goal here is for you to become a "pragmatic translator." This means you translate according to your needs. Sometimes you may just need to summarize or paraphrase some discourse in your natural language; you can use your logic skills to sort out what's being said or the "flow" of some reasoning, and then rewrite it in your own words. Sometimes it is important to assess how groups of compound statements are related in a passage; logic lets you translate to clearly show what kind of compound statements you're working with. If those compounds are related in an *argument*, you'll be able to show if that argument is *valid* (we'll be defining 'argument' and 'valid' a little later in the text). And sometimes arguments are composed of the kinds of sentences where you really need to see all of the logical structure in even "simple" statements—the logical skills you learn here will help you do that as well.

> The more you practice, the better you get.

One further point before we move on to Chapter 2, which introduces our artificial logical language. While one statement can be expressed in many declarative sentences, a single declarative sentence can also be expressing different statements, depending on the context. Consider, for example, the following:

> I will be there next Thursday.

What statement is this expressing? This depends on the speaker, where "there" is, and when exactly next Thursday is. If this were being spoken by a certain president before a famous speech, we might recast this with the more awkward sentence: I (Abraham Lincoln) will be there (in Gettysburg) next Thursday (November 19, 1863). If you, the reader, were to utter the same statement today, the terms "I," "there," and "next Thursday" would have different meanings, and the sentence as a whole would express a different statement. Such terms as these are called *indexicals*. Again, there are some interesting philosophical issues surrounding indexicals that we won't go into in this text. We will, however, show you how to develop translation schemes, which work to provide *context* for your translations from English to the artificial language and vice versa.

Before going on to our artificial language, flex your English-language muscles by doing Exercise 1a.

## Exercise 1a

For each of the following, give 2-3 different English sentences that capture the same meaning as the given sentence. Don't hold back: grab your thesaurus, try to emulate Faulkner or Shakespeare, or hearken back to the slang of your youth.

1. J.K. Rowling is richer than Oprah Winfrey.
2. Cats and dogs are both mammals.
3. Nobody's perfect.
4. Don't we say "thank you" when someone gives us a present?
5. We'll meet for lunch, and you can explain yourself.

6. Her auburn tresses, glittering in the sun, lured an infestation of migrating drones.
7. The professor's garbled precepts did nothing to elucidate the multifarious assignment.
8. John's a doctor and the brother of Bill; Bill is also a doctor and the brother of Andrew; Andrew too is a doctor and the brother of Pete; Pete is the brother of John, Bill and Andrew, and he's also a doctor.
9. Many people want to blend in, try to get along, be on the same wave-length as, or be in harmony with, the prevailing social clique.
10. We use logic every day.

# Chapter Two: Names and Predicates

> *What's Up?*
> LOLA
> What's in a Name?
> Predicates:
> Properties and Relations

As we saw in Chapter One, for any given statement you have multiple ways to express that statement in English. This is also true when translating from one language to another, but in that case, you need to be familiar with *both* languages before you can do such a translation. This chapter introduces you to a new language, our **Lo**gical **La**nguage, which we call **LOLA**.

A natural language, like English, is composed of basic terms for communicating information and a set of rules, called grammar, which tells you how you put those units together. LOLA has similar features. We will introduce basic terms and provide rules governing the proper way to combine these basic terms to produce grammatically correct LOLA sentences.

One important difference between a *natural* language, like English, and an *artificial* language, like LOLA, is that both the terms and the rules are simpler and more precise in the artificial language—something that is both a strength and a weakness of the system. This is not to say that LOLA is somehow automatic or that it won't require you to use your judgment: there's more than one way of successfully translating a given proposition. Furthermore, since one of the goals of creating an artificial language like LOLA is to help us better understand how our own language conveys information, we may focus on different aspects of the proposition when translating an English sentence, depending upon our purposes.

We'll introduce some of the basics of LOLA by considering aspects of the following sentences:

---

**Example Set 2a**
1. Thai is difficult to learn.
2. *Buffy the Vampire Slayer* is great.
3. Last night's sunset was beautiful.
4. Lincoln is a Republican.
5. Bill makes a lot of money.
6. The Middle Ages were depressing.
7. Aristotle is smart.
8. China is big.
9. *War and Peace* is long.
10. Lassie is a collie.
11. Susan paints.

By picking such simple sentences it's easy to identify some common patterns of our language. First, we make distinctions between *things* and their *attributes*. In each sentence, above, there is some *thing*, and there is some *property* attributed to that thing. There is "Thai" or "Aristotle" or "Lincoln" or "*War and Peace*," and there is "is difficult to learn" or "is smart" or "is a Republican" or "is long." This simple distinction is one of the most basic ones we make, and LOLA needs to have a way to represent this distinction.

Another pattern of our language, illustrated in the case of the sentences listed above, is related to our first distinction. "Thing" is admittedly vague and hence, can cover a wide range of terms. We may know a great deal about whatever is being referenced, as in the case of Aristotle or Lincoln or Lassie, or we may not, as in the case of China or Bill. We may be talking about an object, a person, a place, a time period, or a television show. Whatever kind of thing we're talking about and whether we know a great deal about it or not, a basic distinction we often make is between *general* things or specific *individual* ones.

For example, we distinguish between when we're talking about shows or people *in general* versus a *particular* show or person. Think for a moment about the difference between a sentence like #7 that says, "Aristotle is smart" and one that says, "Philosophers are smart." Or consider the difference between sentence #2, "*Buffy the Vampire Slayer* is great" and the sentence, "Horror shows are great." In each case the latter sentence is referencing a group or collective or class, while the sentences from Example Set 2a are all referencing specific individual entities. Again, LOLA will need a way to symbolize this distinction between *individual* and *general* things.

We should attend to one more pattern. Notice that in the above sentences we can't have a proper declarative sentence without both a *thing* and something being said *about* the thing. If you just wrote "Aristotle" or "is fast," we'd say you'd written a sentence fragment. As you learned way back in elementary school, you need both a subject and a predicate to make a complete declarative sentence. Hence, LOLA should not permit either the individual or the attribute to stand by itself; they have to be written together in a grammatically correct LOLA sentence.

To sum up, in order to express the propositional content of the sentences in Example Set 2a in LOLA, we need a way to represent a singular thing, a way to represent the property or attribute being applied to that individual, and a rule that tells us *how* to put them together.

*The symbol that represents the singular thing or individual is called an individual constant*, or simply a *name*. We'll use the lower case letters *a* through *t* of our English alphabet as the pool of symbols that can be used to represent a name. So, "Aristotle" could be represented by the letter "a" or "b" or "e" or "j" or any other letter from a-t, *just so long as the letter has not already been selected as an individual constant for another individual.*

Names: a-t
Predicates: A-Z

There would obviously be enormous confusion if we selected the same letter for two different names.

*The attribute being applied to the individual constant is called a predicate.* We'll use the capital letters A through Z as the pool of symbols for representing predicates. So, "is fast" could be represented as "F" or "M" or "A" or "Y" or any of the capital letters A-Z, again, *only if that letter has not already been selected to stand for another predicate.*

## CONSTRUCTING A TRANSLATION SCHEME

You will be asked to translate from English to LOLA and from LOLA to English, and many times you'll be supplied with what is called a translation scheme. *A translation scheme tells you which letters stand for which names and predicates and any other information necessary to translate a statement from LOLA to English and back.* However, you'll also be asked to construct translation schemes yourself, so we might as well start now. A translation scheme simply lays out what symbols are going to stand for what English terms. Suppose you chose the letter "a" to symbolize "Aristotle." In that case do the following:

1. First, write the letter you've chosen, in lower case
2. Follow the letter you've chosen with a colon
3. Write out the English language name after the colon

**a: Aristotle**

> **EXTRA PRACTICE**
>
> Take a moment to create a translation scheme for each of the individual constants in the sentences in Example Set 2a.

Now we need a translation scheme for the predicate in the sentence "Aristotle is smart." Notice that with the predicate, "is smart," we see another pattern in our language. One of the ways in which a predicate like "is smart" differs from an individual is that the predicate "is smart" can be applied not just to Aristotle, but to any number of individuals. You could apply it to Einstein, Plato, Kant, Feynman, yourself, and many others. So, when we symbolize the property "is smart," we need to have a way to indicate that it doesn't just have to be applied to Aristotle. In short, we need a symbol that stands *in place of any individual constant.* We'll call this place-holder an *individual variable*, or simply a *variable*, and we'll use the lower case letters *u through z* to symbolize it.

A translation scheme for the predicates in Example Set 2a is similar to that for names. If you decide that you want the capital letter S to stand for "is smart," do the following:

1. Write out the capital letter
2. Right next to the capital letter, write down a variable
3. Write a colon
4. After the colon write out the variable you chose, followed by the English language predicate that your symbol represents

**Sx: x is smart**

> **EXTRA PRACTICE**
>
> Take a moment to create a translation scheme for each of the attributes in the sentences in Example Set 2a.

Notice that once you have created the translation scheme, it works both ways. Not only can you symbolize English language statements, but you can take propositions in LOLA and translate them into English.

## WFFS (PRONOUNCED "WOOFS")

Before we move on, we let's look at what we've just created. Is "a" a proposition?

Look back at our definition of proposition: *an assertion that can be true or false*. Is "a" true? Is it false? How would you determine if "Aristotle" is true or false? The question is impossible to answer—*Aristotle is not the kind of thing that can be true or false*. What about "Sx"? Is that true or false? Again, the only response to this question is that Sx is not the kind of thing that can be true or false. That's because, as we said before, a proper declarative sentence in English would have *both* a predicate and some thing that the predicate modifies. So, neither an individual constant nor a predicate all by itself can be a proposition.

*A WFF is a grammatically correct LOLA statement.*

This leads us to a point that has to be kept in mind throughout the rest of the text. There are a variety of ways in which a translation can go wrong. For example, it might be the case that you didn't properly capture the propositional content of a declarative sentence. If instead of "Sx: x is smart" you put down "Sx: x is stupid," then when you translate into LOLA you would have misrepresented our original statement, "Aristotle is smart." Obviously, you need to be accurate and complete in your translation scheme.

However, there is another way you can go wrong. We will introduce new translation rules as we go along, and these must be followed precisely. If you follow the rules, you'll create *grammatically correct LOLA sentences*, which adequately express statements. *A sentence that is properly constructed according to the rules of LOLA is called a well-formed formula.* We'll abbreviate this as "WFF," and pronounce it as, "Woof." If there's a mistake in how our LOLA sentences are put together, we'll say that it's *not* a WFF. It is essential that all our LOLA sentences be WFFs since everything we do depends on this: our building up of progressively more complex statements, our analysis of those statements, and our use of logical techniques to explore the various relationships between and among statements.

> **EXTRA PRACTICE**
>
> Consider the various kinds of mistakes that would create a proposition that's not actually a WFF. Have a classmate deliberately write out some statements that are not WFFs. Take the list, and show why they are wrong and how they could be fixed. Once you've done all of them, switch and do the same thing for your colleague.

## PRACTICING WITH INDIVIDUALS AND ONE-PLACE PREDICATES

Let's put it all together. Take your translation schemes for the predicates and the individuals in Example Set 2a, put them in a box and place it at the top of your document. Below the box, begin to translate from English to LOLA or from LOLA to English. This allows you (or any reader) to quickly reference the scheme as you read through your translations.

### TRANSLATION SCHEME

| t: Thai | b: *Buffy the Vampire Slayer* | s: last night's sunset | l: Lincoln | i: Bill |
|---|---|---|---|---|
| m: the Middle Ages | a: Aristotle | c: China | p: *War and Peace* | e: Lassie |
| n: Susan | Dx: x is difficult to learn | Gx: x is great | Bx: x is beautiful | Rx: is a Republican |
| Mx: x makes a lot of money | Px: x is depressing | Sx: x is smart | Hx: x is big | Lx: x is long |
| Cx: x is a collie | Ax: x paints | | | |

The rule governing how we put names and predicates together is straightforward: *The predicate comes first and is immediately followed by the name it modifies.* So, if you want to symbolize "Aristotle is smart" utilizing the translation scheme that we just constructed, you'd have:

**Sa**

*If you put the individual constant first, it wouldn't be a WFF.* It would be just as if in English you wrote "Aristotle smart is." or "Is smart Aristotle." Finishing up the LOLA translations for Example Set 2a:

1. Dt
2. Gb
3. Bs
4. Rl
5. Mi
6. Pm
7. Sa
8. Hc
9. Lp
10. Ce
11. An

### LOOKING AHEAD

Are there declarative sentences that you think might not be able to be represented by what we've presented thus far? Are there predicates that you think can't be captured by what we've presented yet?

## Exercise 2a

Create a translation scheme for the following sentences, and then translate the sentences into LOLA.

1. Bob moves quickly.
2. Bob wants a great deal.
3. Bob is tall.
4. Susan wants a great deal.
5. Susan moves quickly.
6. Susan is tall.
7. The *Queen Elizabeth* is big.
8. The *Titanic* was tall.
9. The *Titanic* was big.
10. The Roaring Twenties were fun.
11. The Roaring Twenties were dangerous.
12. The White House is small.
13. The White House is grey.
14. Bob is grey.
15. The White House is big.
16. The *Queen Elizabeth* moves quickly.
17. The White House wants a great deal.
18. Bob is fun.
19. Susan is fun.
20. The *Queen Elizabeth* is grey.
21. The *Titanic* moved quickly.

# Exercise 2b

Using the following translation scheme, translate the sentences below from LOLA to English.

## TRANSLATION SCHEME

| f: Frank | s: San Francisco | n: Newark | m: Mary | a: the ambassador | l: Barry Lyndon |
|---|---|---|---|---|---|
| b: Beethoven's Fifth Symphony | Sx: x is quite chic | Cx: x is cold | Dx: x is boring | Lx: x loves a good time | Bx: x is beautiful |

1. Bf
2. Sm
3. Da
4. Db
5. Bs
6. Lf
7. Ds
8. Ls
9. Dn
10. Sa
11. Cf
12. Sn
13. Bl
14. Lm
15. Sm
16. La
17. Sl
18. Ca
19. Ll
20. Ln

## TWO PLACE PREDICATES

Let's consider the following sentence: "Bob is taller than Jim." We *could* represent this as a case of an individual, Bob, with an attribute, "being taller than Jim." However, while we could do this, what we produced wouldn't adequately distinguish between an *attribute* like, "is tall" and the *relation* between one individual (Bob) being "taller than" another individual (Jim). It's one thing to say that "Bob is tall" and it's another to say that "Bob is taller than Jim." After all, we might find out that the first statement is false, but the second statement is true, since Bob might be four feet tall, and Jim only three feet tall.

The statement "Bob is taller than Jim" *compares* two individuals, Bob and Jim. It looks at Bob and Jim in *relation* to each other. A comparative or relational statement means there is more than one thing involved; if you left out one of the items being compared, the proposition wouldn't make sense. Moreover, it means that in order to assess the truth or falsity of the proposition you have to know something about *both* entities, and the relation between the two of them. With this in mind, we need LOLA to capture this language pattern of comparing individuals.

To do this we'll make a distinction between our "attribute" predicates, which we'll call "one-place predicates," and contrast them with "relational" predicates, which we'll call "two-place predicates." The translation scheme for representing two-place predicates is similar to how we construct a one-place scheme, but there are obviously some differences. We still need to assign names to represent the individuals in the original sentence. Now, however, when constructing such a translation scheme, we explicitly state that the predicate is used to capture *a relationship between two things*. For this we need *a capital letter with two place-holding variables*.

So, in constructing a translation scheme for "Bob is taller than Jim," we would get:

**TRANSLATION SCHEME**

| b: Bob | j: Jim | Txy: x is taller than y |

and the symbolization for our original sentence is:

**Tbj**

It's critical when constructing a two-place predicate that you be clear about the *order of the variables*. In our example above, it's the *x* that is taller than the *y*. In symbolizing the sentence, then, you need to be sure that you put the names in the correct relation. If you symbolized our sample sentence as Tjb, this would be read as "Jim is taller than Bob" because, according to our translation scheme the first individual is taller than the second.

## Exercise 2c

Construct a translation scheme for the following sentences, and then symbolize them in LOLA according to your scheme. It may help to first go through and make a list for all the individuals; then list all the predicate relations. Lastly, you can put it all together for your translation scheme.

1. Bobby is smaller than Susan.
2. Susan likes Mary.
3. Detroit is smaller than Newark.
4. Roberto is nicer than Tom.
5. Newark is nicer than Chicago.
6. Mary is uglier than Roberto.
7. Tom is bigger than Roberto.
8. Chicago is smaller than Detroit.
9. Tom is nicer than Mary.
10. Roberto is uglier than Mary.
11. Chicago is larger than Detroit.
12. Mary is jealous of Roberto.
13. Newark is jealous of Detroit.
14. Tom likes Roberto.
15. Tom is jealous of Mary.
16. Roberto likes Tom.
17. Chicago is jealous of Newark.
18. Mary is nicer than Tom.
19. Mary likes Detroit.
20. Tom likes Newark.

## Exercise 2d

Use the following translation scheme to translate the propositions in LOLA into English sentences. When you translate the propositions below you may create a sentence you know to be false. That's fine. Don't change the translation to make it come out true, just translate it exactly as the scheme dictates it should be translated.

**TRANSLATION SCHEME**

| o: one | f: four | n: ten | t: two | i: five |
|---|---|---|---|---|
| l: twelve | h: three | s: six | Exy: x is equal to y | Gxy: x is greater than y |
| Lxy: x is less than y | Sxy: x is succeeded by y | Dxy: x is evenly divisible by y | | |

1. Eoo
2. Got
3. Dnt
4. Ltl
5. Sis
6. Etn
7. Dft
8. Shf
9. Etl
10. Gho
11. Lhn
12. Lol
13. Dsh
14. Sfi
15. Sot
16. Git
17. Dhn
18. Gso
19. Lnl
20. Eon

## Exercise 2e

Create a translation scheme for the following sentences, and then symbolize them in LOLA. Note that some require 1-place predicates, others require 2-place predicates. If you choose you may use the same capital letter for both a one-place and a two-place predicate. For example you could use Sx for "x is a genius" and you could use Sxy for "x is smarter than y." There is no ambiguity—the number of variables tells you whether it's a one-place attribute or a two-place relation.

1. Adam Smith has a snub nose.
2. Plato's nose is bigger than Barbra Streisand's.
3. Barbra Streisand has a snub nose.
4. Groucho Marx is a genius.
5. Karl Marx is smarter than Barbra Streisand.
6. Barbra Streisand is funnier than Plato.
7. Plato is a genius.

8. Adam Smith is funnier than Plato.
9. Barbra Streisand is funnier than Adam Smith.
10. Adam Smith has a bigger nose than Karl Marx.
11. Plato is funnier than Karl Marx.
12. Karl Marx has a snub nose.
13. Barbra Streisand is a genius.
14. Groucho Marx is funnier than Karl Marx.
15. Adam Smith is smarter than Plato.
16. Plato has a bigger nose than Adam Smith.
17. Groucho Marx has a snub nose.
18. Karl Marx is funnier than Adam Smith.
19. Karl Marx is a genius.
20. Barbra Streisand has a bigger nose than Adam Smith.

## Exercise 2f

Using the following translation scheme, translate the propositions in LOLA into English statements. Pay attention to which are two-place predicates vs. one-place predicates.

### TRANSLATION SCHEME

| b: baseball | g: gymnastics | h: hockey | f: football | s: swimming |
|---|---|---|---|---|
| l: lacrosse | Dx: x is deadly | Mx: x is monotonous | Tx: x is a team sport | Vxy: x is more violent than y |
| Bxy: x is more boring than y | Oxy: x is older than y | | | |

1. Vsb
2. Mh
3. Ds
4. Olg
5. Bfl
6. Vgh
7. Dg
8. Ms
9. Tg
10. Vfb
11. Olf
12. Ml
13. Dh
14. Ohb
15. Th
16. Vfl
17. Bsg
18. Db
19. Osf
20. Tl

## THREE-PLACE PREDICATES AND MORE

We can also translate more complex relationships in LOLA. Take the sentence, "John is between Frank and Alice." Again, we might construct a translation scheme in which we talk about John having the property of being between Frank and Alice, but this misses important aspects of these kinds of relationships. As we saw before, one of the features of multi-place relations is the amount of information required to determine whether a statement is true or false. For the claim, "John is seated between Frank and Alice" you need to know about John, Alice, and Frank and their seating arrangement, and what the words "seated between" mean, in order to determine the claim's truth-value.

> **Translation Scheme**
> m: Mary
> l: logic
> d: Disney World
> Lxyz: x learned y at z
> Lmld: Mary learned logic at Disney World

As with two-place predicates, you have to be clear about how the symbolization is representing the relationship. For example, take the equivalent expressions, "Preacher Casey marries Bill to Alice" and "Bill and Alice were married by Preacher Casey." You could construct your translation scheme for the three place predicate "Mxyz" as either "x marries y to z" or "x and y were married by z." Either of these is acceptable, but whichever one you choose, stick to it consistently.

## Exercise 2g

Use the following translation scheme and translate the propositions from LOLA into English sentences.

**TRANSLATION SCHEME**

| c: Canada | s: the United States | f: France | l: Lithuania | g: Greenland |
|---|---|---|---|---|
| Lxyz: x likes y more than z does. | Cxyz: x is closer to y than to z | Txyz: x talks with y more than with z | Fxyz: x is farther from y than z is | Ixyz: x imports more from y than from z |

1. Lcsf
2. Ilfs
3. Fscl
4. Lglc
5. Tlgs
6. Icgf
7. Flgf
8. Tcsf
9. Cglf
10. Lflc
11. Tlcg
12. Igcl
13. Cscg
14. Ffsg
15. Lsfc
16. Tgsf
17. Ifsc
18. Fgsl
19. Clcf
20. Cgsl

## Exercise 2h

Construct a translation scheme for the following sentences, and then translate them into LOLA.

1. Cathy is between Alice and David.
2. The Chicago Museum of Art is between the New York Museum of Modern Art and the Los Angeles Getty Museum.
3. Susan brought David to the New York Museum of Modern Art.
4. Alice introduced Cathy to David.
5. Susan is between David and Alice.
6. David gave the *Mona Lisa* to the Chicago Museum of Art.
7. Susan introduced David to the Chicago Museum of Art.
8. The Los Angeles Getty Museum is between the New York Museum of Modern Art and the Chicago Museum of Art.
9. Cathy gave the *Mona Lisa* to David.
10. David introduced Cathy to Susan.
11. Susan is between Cathy and Alice.
12. Alice brought David to Cathy.
13. Cathy brought Alice to Susan.
14. The Los Angeles Getty Museum gave the *Mona Lisa* to the New York Museum of Modern Art.
15. The *Mona Lisa* is between The New York Museum of Modern Art and the Chicago Museum of Art.
16. Alice brought the *Mona Lisa* to the Los Angeles Getty Museum.
17. Cathy introduced David to the New York Museum of Modern Art.
18. Alice brought David to the Chicago Museum of Art.
19. Susan gave Alice to Susan.
20. Alice gave the Chicago Museum of Art to David.

## Exercise 2i

Use the following translation scheme to translate the propositions from English into LOLA. Notice that there are one, two, and three place predicates.

### TRANSLATION SCHEME

| o: one | t: two | h: three | f: four | i: five |
|---|---|---|---|---|
| s: six | n: ten | l: twelve | Ex: x is even | Ox: x is odd |
| Bxyz: x is between y and z | Gxy: x is greater than y | Sxyz: x succeeds y by z | Dxy: x is evenly divisible by y | Lxyz: x less than y by z |

1. One is between two and three.
2. Four succeeds one by three.
3. Three is odd.
4. Four is even.
5. Four is less than ten by six.
6. Ten is evenly divisible by five.
7. Ten succeeds five by five.
8. Three is between one and twelve.
9. Five is greater than two.
10. Six is less than twelve by six.
11. Four is even divisible by five.
12. Ten is between six and twelve.
13. Five is less than twelve by six.
14. Five succeeds one by four.
15. Three is greater than ten.
16. Two is even.
17. Five is odd.
18. Six is greater than one.
19. Ten is less than twelve by two.
20. Two succeeds one by one.

### LOOKING AHEAD

How would you treat a sentence that makes a claim about all dogs or every grandparent? What would you do about a sentence that just talks about "something" being fast or "some people" being tall?

# Chapter Three: Quantifiers

> *What's Up?*
> Existential Quantifier (∃x)
> Universal Quantifier (x)

In Chapter Two, we worked with grammatically simple sentences about individuals and predicates applied to those individuals, either as attributes or as some kind of relation among individuals. Individuals are perhaps the easiest concept to deal with because we can symbolize them in a straightforward fashion. Since the nature of individuals is that there is only one of them, we don't have to do more than assign one name per individual. Here's an individual; here's a symbol for it. Predicates, on the other hand, are general: they can be applied to more than one individual.

However, there are a number of meaningful sentences that are not reducible to simply an individual and its predicate. Take for example a sentence from a mystery novel: "Somebody murdered the countess." This is a perfectly sensible sentence, and yet we couldn't translate it into LOLA with the system we've constructed so far. 'Somebody' isn't a name, doesn't designate an individual; that sentence designates the fact that there is a non-specified entity (or entities) out there that murdered the countess. We may never learn who that somebody is, but it will still be true that there is *somebody* out there who did the murdering. So, we need some way to represent this state of affairs.

Another reason we need to add to our system is because we don't just apply predicates to individuals, we also apply predicates to some or all *members of classes*. For example, we may want to say not just that "Aristotle is interesting" but "All philosophers are interesting" or "Some philosophers are interesting" or maybe, if we're annoyed, "No philosophers are interesting." Sentences like these can't be represented just by using individual constants and predicates. So, for both these reasons we're going to introduce a new concept, the *quantifier*. There are two of them, and each has its own symbol. We'll begin with the *Existential Quantifier*.

## EXISTENTIAL QUANTIFIERS

Let's start by looking at how to translate a sentence like "Something is red." We'd first decide on how we wanted to represent the predicate. "Is red" is a one-place predicate, so we symbolize it:

### Rx: x is red

> Something is interesting.
> Ix: x is interesting
> (∃x)Ix

Now, we won't move onto representing the individual because we don't *have* an individual to represent. Yet, we can't just leave it at "Rx" since, according to what we said in Chapter Two, this would be a sentence fragment, not a WFF. What we need is a way to say that there is something out there, and we're asserting that this something is red. We need to assert that something *exists* and yet that we aren't going to identify exactly what that thing is. This is what the existential quantifier does.

An existential quantifier looks like this: ∃. It is enclosed within parentheses ( ) and it is always accompanied by a single individual variable (remember, that's any lower case letter from u-z), so we can represent it as any one of the following: (∃u), (∃v), (∃w), (∃x), (∃y), or (∃z). It goes *in front of* the predicate in your symbolization, and the predicate is followed by that same individual variable. This gives you:

### (∃x)Rx

You'd read this, in "logic speak" as "There exists some x, such that x is red" or in English as "Something is red." As we said a moment ago, we're actually *asserting two things simultaneously*: that something exists, *and* that *this* something happens to have the following particular property or attribute. Hence, to be grammatically correct, the variable you select to put beside the existential quantifier *must be the same letter you choose to put beside the predicate*. So, you would *not* write (∃**y**)R**z**—this is not a WFF. *The variable placed beside the predicate is said to be "bound" by the quantifier. Variables appearing in WFFs are always bound variables.*

The claim "Something is red" is not specific about how many things are red. All the proposition (∃x)Rx is saying is that "There is *at least one* red thing." This means that the existential quantifier is extremely versatile. We can use it to represent "some," "many," "a few," "lots and lots," "most," "several," and similar expressions. *Any English term that asserts that at least one thing exists can be represented by the existential quantifier.*

It may strike you as odd that the two sentences, "Some things are red" and "Most things are red" are *both* represented by the same symbolization, (∃y)Ry [or (∃x)Rx or (∃z)Rz, depending upon what individual variable you select; all these sentences mean the same thing]. After all, there is a difference between the claims, in English, that *some* things are red and the claim that *most* things are red. This you may view as either a limitation or a strength of LOLA. English, unlike LOLA, is meaning-rich. You can be extremely nuanced and subtle in English. But remember, the main purpose of developing LOLA is to adequately represent the *logical* structure of English statements. The existential quantifier adequately represents that we are talking about some existing members of a class.

## MIXING EXISTENTIAL QUANTIFIERS AND INDIVIDUALS

Just as you can have relationships between two individuals, such as "Bill is funnier than Sam," you can have a relationship between an individual and a "something," as in the sentence, "Something is funnier than Sam." We'll learn how to do this by considering the following example set.

> Mary loves something.
> m: Mary
> Lxy: x loves y
> (∃x)Lmx

### Example Set 3a
1. Something is taller than Sam.
2. Sam is taller than something.
3. At least one thing is more beautiful than the Mona Lisa.
4. There are many things more beautiful than Sam.
5. Sam gave something to Lisa.

Sentences 1 and 2 both include an individual, "Sam," and a term that requires the use of an existential quantifier, "something." Just as in the second chapter, we need to create an individual constant to represent the individual "Sam" and we'll use lower case "s"—**s: Sam**.

The sentences also both employ the two-place predicate "taller than" which can be symbolized:

**Txy: x is taller than y**

Now you're set to symbolize the first two sentences.

When you use a quantifier, it goes at the beginning of the WFF, even if the sentence includes individuals. So, in this case you would put down the left parentheses, put in the ∃, select a variable and write it down to the right of the ∃, and then close the parentheses. This results in:

**(∃y)**

Then put down the predicate symbol:

$$(\exists y)T$$

Next, you look at your original sentence, and ask yourself, just as you did before when you were first working with two place predicates, "What is the relationship between the two items?" For example, sentence 1 would have to be symbolized as:

1. **$(\exists y)Tys$**

because sentence 1 states that *something* is taller than Sam. However, sentence 2 would be symbolized as:

2. **$(\exists y)Tsy$**

because it states that Sam is taller than something.

Now consider sentences 3 and 4. Keep the same individual constant for "Sam" and translate "more beautiful than" and "the Mona Lisa":

**Bxy:** x is more beautiful than y     **m:** Mona Lisa

This results in:

3. **$(\exists x)Bxm$**
4. **$(\exists x)Bxs$**

Notice that even though one English sentence says "many things" and the other says, "at least one thing" they are both represented with the existential quantifier. We lose this distinction in this basic translation into LOLA.

Finally, sentence 5 has a three-place predicate, two individuals, and a term that requires an existential quantifier. Assuming that you keep the same representation for Sam, and symbolize "Lisa" and the "gives to" relationship, respectively, as:

**l: Lisa**
**Gxyz: x gives y to z**

Sentence 5 would then be translated as:

5. **$(\exists z)Gszl$**

## LOOKING AHEAD

Consider what you might do if all you were talking about was different kinds of cars: fast cars, blue cars, sports cars, etc. Would there be any way to set up your translation scheme to address this fact?

## RANGE OF DISCOURSE

Before we move on to practicing this new kind of symbolization, we need to take a moment to attend to the *kinds* of entities covered by a quantifier. Consider the difference between sentences like "Some*thing* is tall" or "Some*thing* is fast" and those like "Some*one* is tall" and "Some*one* is fast." Think for a moment about how you would translate them and whether you could use a single translation scheme for all four. You'll see that this wouldn't be possible, as of yet, because even though in both cases you're making general claims, the former sentences are *more* general than the latter. "Someone," which refers to some *person*, is actually a subcategory of "something," thing being the widest class of entities we deal with.

> Somebody loves Mary.
> RD: People
> m: Mary
> Lxy: x loves y
> (∃x)Lxm

This raises an interesting point about the pragmatic nature of logic and reasoning. Often, in order to achieve greater clarity about a subject, we narrow our inquiry so that it only includes those elements we believe are directly pertinent to the subject. If our subject is automobile engines, we may not be concerned at all with how they interact with Renaissance paintings or chalkboards or people or anything *but* various kinds of car engines. If we are looking at plants we may well not care at all about anything *but* plants for the purposes of an investigation.

As we narrow a field of inquiry or area of interest, we also *restrict our discourse*. If we are talking about plants, say, and our listeners or readers know this, we don't need to keep repeating the word 'plant,' we can use the pronouns 'it' and 'they,' and everyone understands that these terms refer to plants (rather that Puritans or Presidents). Sometimes we may not even be aware of this because we "automatically" supply the antecedents of pronouns, because we know the context.

In LOLA, however, we must explicitly designate a context in our translation scheme; we can state how "wide ranging" the subject matter of the discourse is. This allows any reader of LOLA translations to know exactly what the range of discourse is, and to quickly and accurately read off the translation. For example, if we knew that Bx: x is beautiful, and we knew that our range of discourse was limited to plants, we would translate (∃x)Bx as "Some *plants* are beautiful" rather than "Some*thing* is beautiful."

So, we'll add a new component to our system that allows us to do this. From now on, when a translation scheme is presented to you, it will include an explicit indication of the range of the discourse (RD). Often the statements will refer to the widest general range of subjects, and we will list the RD as "everything." Similarly, when *you* construct a translation scheme you will need to specify the RD. On many occasions, the RD will be "everything" because the sentences cannot be restricted to some particular domain. However, you should be alert to those occasions in which it would be reasonable to restrict the discourse—for the very practical reason that it will often simplify your translations, without sacrificing clarity. Let's illustrate this with the following sentences:

**Example Set 3b**
1. Some trees are pretty.
2. At least one tree is short.
3. Most trees are tall.
4. The tree in my front lawn is bigger than most trees.
5. Some trees are fuller than the one in my back yard.

In the "discourse" of Example Set 3b, all the sentences are about trees. So, you could construct the following translation scheme:

**TRANSLATION SCHEME**
RD: Trees

| l: the tree in my front yard | b: the tree in my back yard | Bxy: x is bigger than y | Fxy: x is fuller than y | Sx: x is short |
|---|---|---|---|---|
| Px: x is pretty | Tx: x is tall | | | |

This would then allow you to symbolize the sentences in Example Set 3b as follows:

1. (∃x)Px
2. (∃y)Sy
3. (∃z)Tz
4. (∃v)Blv
5. (∃w)Fwb

Or consider the sentences in the following set:

**Example Set 3c**
1. One is greater than some number.
2. Some number is between two and six.
3. At least one number is equal to zero.
4. One hundred is a large number.
5. Five is smaller than lots of numbers.

Since all sentences in Example Set 3c are about numbers, you could construct the following translation scheme:

**TRANSLATION SCHEME**
RD: Numbers

| o: one | h: one hundred | r: zero | f: five |
|---|---|---|---|
| t: two | s: six | Lx: x is large | Gxy: x is greater than y |
| | | Exy: x is equal to y | |
| Bxyz: x is between y and z | Sxy: x is smaller than y | | |

This would result in the following symbolizations:

1. (∃x)Gox
2. (∃w)Bwts
3. (∃x)Exr
4. Lh
5. (∃z)Sfz

Although the concept of Range of Discourse may seem odd at first, remember what we said about how we talk in natural language. If, as above, our conversation is about numbers, we wouldn't keep saying "number" over and over again. We'd use pronouns: "it" and "they." We can think of the RD acting like the *antecedent of a pronoun*: the term that gives the pronoun meaning. Our bound variables—those variables attached to quantifiers—are like pronouns, they get their meaning from the RD. For sentence 1

RD: Numbers
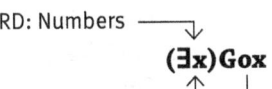
**(∃x)Gox**

We read: There exists a number such that one is greater than **it** is.
Or simply: One is greater than some number.

Because the RD for a group of statements limits what those statements talk about, any individual constants in those statements must name items in that RD. And remember, when there are no restrictions on what is being discussed in the statement sets, the RD will be "Everything."

## Exercise 3a

This exercise focuses on existential quantifiers and the RD. But, it also includes sentences that don't make use of existential quantifiers. So, be alert! We've broken the exercise into smaller units to make constructing the appropriate RD easier. We suggest you use the following method to design your translation scheme. It draws on the process from the previous chapter, and includes the new elements as well.

- Look through the sentences in the set and determine the appropriate RD.
- Identify all the individuals and assign them individual constants.
- Identify all the predicates, determine if they are one, two, or three place predicates, and then construct the proper symbolization for each predicate.

Construct translation schemes for the following sets, and then symbolize the sentences in LOLA.

*Set 1*
1. Somebody is bored.
2. Many people like to travel.
3. Bob likes to fly more than some people.
4. Susan is in danger.
5. At least one person likes to fly more than Susan.
6. Some people are in danger.
7. Bob likes to fly more than Susan.
8. Susan likes to fly more than Bob.
9. Bob is bored.
10. Bob is in danger.

*Set 2*
1. Some things go fast.
2. Most things are physical objects.
3. Air Force One uses more energy than some stuff.
4. Some things are physical.
5. My bike transfers energy to at least one entity.
6. A few things go faster than Air Force One.
7. My bike goes faster than at least one thing.
8. Air Force One goes fast.
9. Something transfers energy to Air Force One.
10. My bike goes faster than Air Force One.

*Set 3*
1. Some animals are mammals.
2. Most animals eat meat.
3. Sam is a mammal.
4. At least one animal is furry.
5. A few animals are furrier than Rover.
6. Sam carries food to some animals.
7. Rover is furrier than Sam.
8. Sam eats meat.
9. Some animal carries food to Rover.
10. Rover is furrier than some animal.

# Exercise 3b

Using the following translation schemes, translate from LOLA to English.

## Set 1
**TRANSLATION SCHEME**
RD: Sub-atomic particles

| l: lepton 'a' | e: electron 'b' | p: proton 'c' |
|---|---|---|
| Dx: x disintegrates | Px: is polarized | Ixy: x ionizes y |
| Cxyz: x causes y to run into z | | |

1. $(\exists x)Px$
2. $(\exists y)Iyp$
3. $(\exists x)Dx$
4. $(\exists z)Clpz$
5. Iel
6. $(\exists x)Iex$
7. Pl
8. $(\exists x)Ipx$
9. $(\exists y)Cype$
10. Dl

## Set 2
**TRANSLATION SCHEME**
RD: People

| m: Mary | b: Bob |
|---|---|
| k: Karen | Wx: x whines |
| Lxy: x likes y | Txyz: x telephones y about z |

1. $(\exists x)Txmk$
2. Wm
3. $(\exists z)Lbz$
4. $(\exists y)Tmby$
5. Lmb
6. $(\exists x)Wx$
7. $(\exists x)Lxm$
8. $(\exists z)Tzkb$
9. Tmbk
10. $(\exists x)Lbx$

## Set 3
**TRANSLATION SCHEME**
RD: Everything

| g: the IGA grocery store | j: John |
|---|---|
| h: John's house | Rx: x is run-down |
| Sxy: x is south of Y | Bxyz: x is between y and z |

1. Rg
2. $(\exists x)Rx$
3. $(\exists x)Sxh$
4. Bjgh
5. $(\exists z)Bjgz$
6. $(\exists w)Sgw$
7. $(\exists x)Bgxj$
8. Sjh
9. Bghj
10. $(\exists z)Sjz$

> **THINKING AHEAD**
>
> How would you handle a statement about *all* people? Or *all* dogs? Or *everything* for that matter? Would you be able to do it with what we have in LOLA thus far? If not, what kind of symbols or rules would you need to adequately represent such a sentence?

## UNIVERSAL QUANTIFIERS

Judy trusts everybody.
RD: People
j: Judy
Txy: x trusts y
(x)Tjx

There is one more quantifier that we'll be using in LOLA and it is called the *universal* quantifier. As its name suggests, it is used to express the notion of "every" or "all," as in the sentences, "Everything is made of water" or "All things are in motion." *It is symbolized by using one of the individual variables (lower case letters u-z) enclosed in parentheses: (x), (y), (z), (w), (v), (u).*

Assuming the property "made of water" is symbolized as Wx: x is made of water, the sentence "Everything is made of water" would be constructed in the following manner:

1. Write down the quantifier just as you did before, although in this case, it's a universal quantifier: **(x)**
2. Write down the predicate symbol: **(x)W**
3. Write down the same letter that was used for the individual variable in the universal quantifier: **(x)Wx**

You would read this as, "For all (any) x, x is made of water." Or, more simply, "Everything is made of water."

Much of what we covered in the first part of this chapter regarding existential quantifiers will apply to universal quantifiers, although now there are even more possible combinations. We can have statements with just universal quantifiers, with individuals and universal quantifiers, with universal quantifiers and existential quantifiers, and with universal quantifiers, existential quantifiers, and individuals. Moreover, we can have statements with more than one existential quantifier, or with more than one universal quantifier. We have put together some of these variations in the next example set.

## Example Set 3d
1. All people struggle for freedom.
2. Anyone can do it.
3. Everyone wants a good career.
4. Each person changes throughout their life.
5. Everybody loves Raymond.
6. John loves people.
7. Bob makes everyone laugh.
8. John told everybody about Bob.
9. Bob told John about everybody.
10. Everyone told John about Bob.

We'll begin by constructing a translation scheme for the set of sentences. Since all the sentences deal with people we can restrict our discourse to people. Our next step is to identify all the individuals and assign them individual constants. We then look at all the predicates, see whether they are one, two, or three place predicates, and then design an appropriate set of symbolizations for them. This gives us:

## TRANSLATION SCHEME
RD: People

| j: John | b: Bob | r: Raymond | Cx: x can do it | Fx: x struggles for freedom |
|---|---|---|---|---|
| Wx: x wants a good career | Tx: x changes throughout their life | Lxy: x loves y | Hxy: x makes y laugh | Gxyz: x told y about z |

Sentences 1, 2, 3, and 4 are all similar to the examples we just did.

1. **(z)Fz**
2. **(y)Cy**
3. **(x)Wx**
4. **(w)Tw**

In sentence number 5 we have a two place predicate, which expresses a relationship between an individual "Raymond" and a term that would require a universal quantifier to represent it, "everybody." Just as we did in the case of the existential quantifiers, we begin with the quantifier:

**(x)**

Then we put down the predicate term:

**(x)L**

Next we put down the remaining terms in appropriate order:

$$5.\ (x)Lxr$$

This, read in logic-speak, would be, "For any person, that person loves Raymond."

Remember that symbolization is not a mechanical process, and you have to be alert to the different ways a statement can be expressed in English. We could say that "Raymond is loved by everybody." This is just another English alternative of our original statement "Everybody loves Raymond." So even though 'Raymond' and 'everybody' "changes places" in the English sentence, both sentences express the same proposition, and so would be translated exactly the same in LOLA: (x)Lxr.

To further clarify this, let's look at sentence 6 and see how it differs from sentence 5.

Again, we begin with a universal quantifier, and the predicate. However, this time instead of everyone loving the individual, the individual loves everyone. This means that the order of the individual constant and the variable will be switched, resulting in:

$$6.\ (x)Ljx$$

Take a moment and consider how you would symbolize sentence 7. Ask yourself what the relationship is, what your translation scheme gives you, and then give it a try.

You should have gotten:

$$7.\ (z)Hbz$$

In sentence 8 we have a universal quantifier, and a three-place predicate. The order remains the same: quantifier, predicate, and then the appropriate ordering of the individuals and variable. This results in,

$$8.\ (x)Gjxb$$

The last two are variations on the same relationship and terms, so it's just a matter of putting them in correct order. Try it by yourself first, and see if you get them correct.

You should have written the following:

$$9.\ (y)Gbjy$$
$$10.\ (v)Gvjb$$

## USING MORE THAN ONE QUANTIFIER

Let's now consider how we would symbolize the following sentence,

> Somebody trusts everybody.
> RD: People
> Txy: x trusts y
> (∃x)(y)Txy

1. Everything influences everything.

We have a two-place predicate, but no individuals at all, only two quantifiers. While this raises new issues, we should begin with what is the same from our earlier work. First, we specify the RD, which in this case would be everything. Then we will construct a representation for the expression "influences":

**Ixy: x influences y**

Now, how to symbolize the rest? Since the sentence uses *two* expressions referenced by universal quantifiers, you will need *two* universal quantifiers in your symbolization. You can't pick the same quantifier for each one of these terms as that would result in confusion. This isn't much of a problem since you have six different variables to choose from (u-z). So, select a variable to be the first universal quantifier and write it down:

**(x)**

Then select a different variable to represent the second universal quantifier, and write it down to the right of the first one. This gives you:

**(x)(y)**

Now you can write down the predicate to the right of the last quantifier, giving you:

**(x)(y)I**

and now you need to write down the individual variables that match up with the quantifiers. This gives you:

**(x)(y)Ixy**

And you're done!
   Let's now consider the following two sentences

2. Everything influences something.
3. Something influences everything.

These are two very different claims, and we need to be able to represent the differences symbolically. Begin with sentence 2, and look at the following two symbolizations

$$(x)(\exists y)Ixy \qquad (\exists x)(y)Ixy$$

The one on the left says, "For any x, there exists something y, such that x influences y." The one on the right says, "There exists something x, such that for anything y, x influences y." Now ask yourself which one best represents sentence 2, and you'll probably agree that it's the one on the left. Turn to sentence 3 and try the same thing and you'll find that Sentence 3 is best rendered with the expression on the right.

If you're using two or more different quantifiers, you'll need to make sure you put them in the proper order. On the other hand, if the quantifiers are of the *same* type, then it may not make a difference. For example, in the sentence above "Everything influences everything" it wouldn't matter whether you put the "x" or the "y" first. Any of the following sentences would convey the exact same meaning.

$$(x)(y)Ixy \qquad (y)(x)Ixy \qquad (x)(y)Iyx \qquad (y)(x)Iyx$$

The main point is simply to pay close attention to what exactly the sentence says, and what the translation scheme says, and that the order of the quantifiers and individual variables takes all this into account. We have already seen with the notion of predicates that it matters a great deal how exactly you phrase the translation scheme; "x influences y" is not the same as "x is influenced by y." These small variations make for big differences.

Finally, we will briefly examine a situation in which you would use three quantifiers. All the same issues still obtain, but keep in mind that these need even more care in translation, especially regarding the order of the quantifiers.

Assume the following:

**RD: Numbers**
**Bxyz: x is between y and z**

Think about how to translate the following sentences. In each case we have supplied you with two different English language sentences expressing the same proposition.

1a. Some number is between some number and every number.
1b. Between every number and some number is some number.

2a. Every number is between at least one number and every number.
2b. Between every number and some number lies every number.

3a. Every number is between some number and at least one number.
3b. Between at least one number and some number is every number.

4a. Some number is between every number and some number.
4b. Between every number and some number exists some number.

Here are the symbolizations:

1. (∃x)(∃y)(z)Bxyz
2. (x)(∃y)(z)Bxyz
3. (x)(∃y)(∃z)Bxyz
4. (∃x)(y)(∃z)Bxyz

We conclude this chapter with practice exercises that draw on everything you've learned thus far in the first three chapters.

## Exercise 3c

*Set 1: Translate from English into LOLA*
**TRANSLATION SCHEME**
RD: Animals

| l: Leo | c: Casper | r: Rover |
|---|---|---|
| Tx: x is timid | Hxy: x hisses at y | Axy: x is afraid of y |
| Sxyz: x scares y more than z | | |

1. Every animal is afraid of some animal.
2. Leo hisses at Casper.
3. Casper is timid.
4. Casper scares some animals more than Rover.
5. At least one animal is timid.
6. All animals scare Rover more than Casper.
7. At least one animal scares every animal more than Rover.
8. Rover scares Leo more than Casper.
9. Most animals are afraid of some animal.
10. Most animals hiss at Casper.

## Set 2: Translate from LOLA into English
**TRANSLATION SCHEME**
RD: Numbers

| n: zero | g: a google | Ox: x is odd |
|---|---|---|
| Mxy: x is more than y | Ixy: x is identical to y | Dxyz: x divided by y results in z |

1. (∃x)Mxg
2. (∃y)Oy
3. (∃z)(x)Mzx
4. Inn
5. (x)(y)(∃z)Dxyz
6. (x)(∃y)Dxyn
7. (x)(∃y)Ixy
8. Og
9. (x)Mxn
10. (z)Dznn

## Set 3: Translate from English into LOLA
**TRANSLATION SCHEME**
RD: Philosophers

| a: Aristotle | s: Aristotle's son | k: Kant |
|---|---|---|
| Wx: x is wise | Sxy: x is smarter than y | Txy: x teaches logic to y |
| Pxyz: x passes his wisdom along to y via z | | |

1. All philosophers are wise.
2. One philosopher is smarter than all philosophers.
3. All philosophers teach logic to some philosophers.
4. Aristotle teaches logic to all philosophers.
5. All philosophers pass their wisdom along to all philosophers via some philosopher.
6. Aristotle passed his wisdom along to all philosophers via his son.
7. Aristotle is wise.
8. Kant is smarter than at least one philosopher.
9. Kant teaches logic to himself.
10. All philosophers are smarter than at least one philosopher.

*Set 4: Translate from LOLA into English*
**TRANSLATION SCHEME**
RD: People

| s: my sister | m: my mom | b: my brother |
|---|---|---|
| Nx: x nags | Fxy: x fights with y | Lxyz: x wants y to love z |

1. (y)Ny
2. (∃y)Ny
3. (∃x)Lmxs
4. (x)(y)(∃z)Lxyz
5. Lmsb
6. (∃y)(∃w)Fyw
7. (x)(y)Fxy
8. (∃x)(y)Fxy
9. (y)(∃z)Fyz
10. (∃z)Lszb

# Exercise 3d

Construct your own translation scheme and then symbolize the following sentences.

1. Any car is faster than Dan's Suzuki.
2. Lane's Honda can pass most cars.
3. Most cars cost more than Dan's Suzuki.
4. The President's limo is parked between Dan's Suzuki and Lane's Honda.
5. Lane's Honda is pretty.
6. The President's limo is faster than every car.
7. At least one car is parked between the President's limo and Dan's Suzuki.
8. Some cars are pretty.
9. At least one car is faster than Lane's Honda.
10. At least one car can pass every car.

### THINKING AHEAD
Think about a sentence that has an "and" in it or "but" or "in addition." How does that change the sentence? Consider how our grammar allows us to combine various statements into larger compound statements. Can LOLA do this?

# Chapter Four: Negations and Conjunctions

> **What's Up?**
> Not ~
> And &

In Chapters Two and Three we introduced you to predicates, individuals, and quantifiers, and showed you how these tools can be used to represent a large number of propositions. These, however, will only let you symbolize a small percentage of the total possible statements that we commonly use. In this chapter we provide you with additional tools that expand the range of statements you can express in LOLA.

## THE TILDE
Consider the following expression:

1. The *Mona Lisa* is lovely. (m: *Mona Lisa;* Lx: x is lovely)

You should have no problem symbolizing this as: Lm
But what if you wanted to say:

2. The *Mona Lisa* is *not* lovely.

How would you go about expressing this proposition? One possibility of course is to say there is a property, *non-lovely*, and the *Mona Lisa* possesses such a property. You would then create a translation scheme to represent this property and proceed as usual.

But this seems unsatisfactory. In many cases we don't talk about entities possessing *non*properties, we talk about them *lacking* particular properties. If something is lacking the property loveliness, then we would not assign the predicate Lx: x is lovely to it. Looked at in this way, LOLA needs a way of representing that the *Mona Lisa* should not have the predicate "is lovely" assigned to it.

Also consider that in sentence 1 a claim is being made, namely that the *Mona Lisa* is lovely. So, what is sentence 2 doing? A reasonable way to look at it is to say that *it is a claim about the truth status of sentence 1, namely a claim that sentence 1 is false*. What we have then

> Logic isn't complicated.

is a disagreement about the truth status of sentence 1. This is obviously an enormously important pattern in our reasoning process. We don't just make claims, we make claims and counter claims. So, we should have a way of expressing that many times the propositions we assert are actually propositions *about other propositions*. This means we need some way of expressing this aspect of our language.

For these reasons, we need to introduce a new element into LOLA. This new element will be called **negation** and the symbol we will use is called the **tilde** and it looks like this: ~. The tilde is used in representing such English terms as: *not, it's false, un-, isn't the case, is not true*, and so on. Negation means that we are asserting that some statement is not true.

We'll demonstrate how to symbolize a statement using the tilde by considering the sentences in Example Set 4a. As you'll see, the placement of the tilde in a proposition depends a great deal on what you're saying.

### Example Set 4a

**TRANSLATION SCHEME**
RD: People

| b: Bob | s: Sam | j: Janice |
|---|---|---|
| Tx: x is tall | Lxy: x likes y | Gxyz: x is telling the truth about y when talking to z |

1. Bob isn't tall.
2. Bob doesn't like Sam.
3. Bob is not telling the truth about Sam when talking to Janice.

Notice that in the translation scheme, there are no negations; everything is expressed in positive terms: *x likes y, x is tall, x is telling the truth about y to z*.

The first thing to remember about negation is that it acts *on a proposition*. In other words, you have to have a proposition first, and *then* you negate it. With this in mind, consider sentence 1. It is saying that "it is false that Bob is tall." An easy way to begin is by simply symbolizing the positive part of the sentence, that is, the part after "it is false." If you do that you get:

$$Tb$$

Now we need to say that this statement is false, that it's *not* the case that Bob is tall. In LOLA, the tilde functions a bit like the quantifier, it comes *before the proposition you want to negate*. If you put the tilde between the "T" and the "b" then you'd have a *non*-WFF. One way to think of this is to imagine the negation as standing for the phrase, "It is not the case that ...". So, you'd write down your tilde to the left of the proposition:

**1. ~Tb**

You can read this as "It's not the case that Bob is tall" or "It's false that Bob is tall." Try the same procedure with sentence 2. Symbolize the positive claim, "Bob likes Sam" as:

**Lbs**

Then deny this claim by placing the tilde to the left of it. This gives you:

**2. ~Lbs**

Now try sentence 3. The positive component of sentence 3 is "Bob is telling the truth about Sam when talking to Janice" which is symbolized Gbsj. Again, place the tilde to the left of it, which gives you:

**3. ~Gbsj**

One final note on negation: you can put two or more tildes into a proposition. Take for example the sentence, "That was not an unfair play that Michael made" (m: Michael; Fx: x makes a fair play). Whether or not you believe the sentence is in good writing style, the "not ... un" construction is used frequently in both our spoken and written language, and we should be able to capture this in LOLA. And we can. Our sentence above can be analyzed as a positive claim that is denied twice, and we'd symbolize it as:

**~~Fm**

One final word of caution on translation. Remember that you are a native speaker of the English language, so don't approach symbolization in a mechanical manner. For example, consider the sentence: Bob didn't unpack my car. According to our previous example, you might want to symbolize this as: ~~Pbc (Pxy: x packs y; b: Bob; c: my car). Or, "It's false that Bob didn't pack my car." This means, of course, that Bob did pack my car. But, as native speakers, we know that "*not* packing" is different than "*un*packing." So a better translation scheme would be:

b: Bob    c: my car    Uxy: x unpacks y

We'd symbolize the statement "Bob didn't unpack my car" as

~Ubc

## Exercise 4a

Symbolize sentences 1-10 in LOLA, and then translate propositions 11-20 into English sentences. Use the following translation scheme.

**TRANSLATION SCHEME**
RD: Everything

| s: the sun | e: the earth | m: the moon | p: Pluto |
|---|---|---|---|
| j: Jupiter | Gx: x is made of green cheese | Hx: it's hot on x | Lx: x can support life |
| Fxy: x is friendlier than y | Txy: x has more traffic jams than y | Cxyz: x causes y to affect z's orbit | |

1. The earth is not made of green cheese.
2. The moon is made of green cheese.
3. The sun isn't friendlier than the earth.
4. Jupiter isn't friendlier than the sun.
5. The moon has more traffic jams than the sun.
6. It's not un-hot on the sun.
7. The sun causes the moon to affect the earth's orbit.
8. The moon doesn't cause the sun to affect the earth's orbit.
9. The earth doesn't cause the moon to affect the sun's orbit.
10. The sun can't support life.
11. ~Gp
12. ~Hp
13. Tes
14. ~Tej
15. ~Cjse
16. Fjm
17. ~Le
18. Cspj
19. ~Fpj
20. ~Hs

# CHAPTER FOUR: NEGATIONS AND CONJUNCTIONS

> **EXTRA PRACTICE**
>
> Take sentences 11–20, construct your own translation scheme for them, and translate them into English.

## Exercise 4b

Construct a translation scheme for the following sentences, and then symbolize them accordingly.

1. Detroit is dull.
2. Chicago is not close to Los Angeles.
3. Los Angeles isn't between Chicago and New York.
4. Detroit has lots of factories.
5. Los Angeles isn't close to New York.
6. New York doesn't have lots of factories.
7. Chicago is between Detroit and Los Angeles.
8. Chicago isn't dull.
9. Los Angeles is close to Detroit.
10. Chicago isn't between Detroit and New York.

> **THINKING AHEAD**
>
> How would you treat a sentence like, "Somebody doesn't like Joe"? or "Nothing is physical"? Do we need to have a new symbol? Can we do it with what we already have in LOLA?

## A TILDE AND QUANTIFIER TOGETHER

The first set was relatively straightforward since it only involved individuals and predicate relationships. Now let's look at how we would deal with tildes and quantifiers together.

> Somebody doesn't like Raymond.
> $(\exists x)\sim Lxr$

Unfortunately, it won't be possible to use the same simple method we did in the previous section. To see why consider the following two sentences:

1. Some people don't like Bob.
2. There is not a single person out there who likes Bob.
   (RD: People; b: Bob; Lxy: x likes y)

Sentence 1 is asserting that *some people* out there have a particular relationship with Bob; namely, they don't like him. Begin by writing down an existential quantifier thereby saying that something exists:

$$(\exists x)$$

Now, ask yourself, what is it that exists? The sentence is saying that what exists are people who don't like Bob. So the next step is translating this part of the statement. Begin by looking at the statement as though it were a positive assertion, "x likes Bob" and symbolize that. This gives you

$$Lxb$$

Now, you want to claim that this is false. You want to say that it's not the case that x likes Bob, so you put a tilde in front of the statement, resulting in:

$$\sim Lxb$$

Now place all of this to the right of your existential quantifier and you get:

**1. $(\exists x)\sim Lxb$**

This says, "There exists some person, such that that person does not like Bob" or more colloquially, "Somebody doesn't like Bob," or "Some people don't like Bob."

Contrast this with sentence 2, "There is not a single person out there who likes Bob." In this case the sentence is saying that *it's false that there is even one person out there who does like Bob*. Again, it helps to begin by symbolizing the statement that you will then claim is false. Ask yourself how you would symbolize the sentence "There is a person out there who likes Bob." You'd begin with your existential quantifier:

$$(\exists v)$$

Then you'd complete it by symbolizing the phrase, "v likes Bob" and put that to the right of the quantifier, resulting in:

$$(\exists v)Lvb$$

Now, *this* is the assertion that you want to claim is false. You would therefore finish up by placing the tilde to the left of the statement and have:

**2. $\sim(\exists v)Lvb$**

This, when read aloud, would be, "It's false that there exists even one person who likes Bob."

You use the same basic approach when it comes to statements that involve both a tilde and a universal quantifier. Take sentence 3, for example

3. Everyone dislikes Bob.

Ask yourself what *exactly* is the claim here? It is stating that for *any* individual you pick they are *not* going to *like* **Bob**. What is being negated is the "liking statement" NOT the universal statement. The correct symbolization, then, is:

3. (x)~Lxb

## OPERATORS AND SCOPE

Before we continue with symbolization we need to introduce a new piece of terminology, **operators**. As we pointed out in section one, negations are claims *about* other claims. Another way to say this is that the tilde *operates* over the statement. A quantifier also operates over a statement in that it says how many of a particular entity (some or all). Hence, we also call them operators. This means we thus far have three different operators:

**Existential Quantifiers:** (∃u), (∃v), (∃w), (∃x), (∃y), and (∃z)
**Universal Quantifiers:** (u), (v), (w), (x), (y), and (z)
**Negation:** ~

There will be more operators introduced later, but these first three will be enough to give you the idea.

Obviously, not all statements have operators since a statement like "Bob likes Susan" has neither quantifiers nor tildes in it. However, an important feature about operators is that while there can be a number of them in any given proposition, there will always be *one and only one main operator*. In a moment we will show you how to recognize what the main operator of a proposition is, and then you will be expected to be able to identify the main operator for any given statement. *Any symbolization in which it is impossible to definitely identify the main operator means it is a non-WFF in LOLA.*

By comparing these two symbolizations we can now state the rule for determining the main operator of a statement.

(∃x)~Lxb    ~(∃v)Lvb

We'll start with the one on the left.

Begin by identifying all the operators in the statement. You'll see that there are two: the existential quantifier, and the tilde. Pick either one of these operators and ask yourself, what part of the formula is it *operating on*? The tilde is saying that the relationship of x liking Bob is false. It is "operating on" the claim about the relationship of Bob and x. However, the tilde is not making any claims about whether the quantifier is true or not. In other words it is not operating on the quantifier. *The extent of the formula covered by the operator is called its scope.* On the other hand, the existential quantifier is making a larger claim. It is asserting that there exists a person about whom it is false to claim that they have a particular relationship with Bob. We say that the quantifier has a *wider* scope than the tilde does because it takes in more of the formula than the tilde, and this makes it the main operator.

Another way to think of this is to ask yourself, is the focus of the sentence the proposition that something is false or is the focus that something exists? If it's the former, then the tilde will be the main operator, and if it is the latter then the existential quantifier will be the main operator.

Now, turn to the formula on the right. Again, we have two operators: one a tilde, the other an existential quantifier. So, again ask yourself the extent of the formula that is being addressed by the operator and compare the two. The quantifier is asserting that there exists a person and that this person bears a particular relationship to Bob, namely, he likes him. The tilde operates over all that; it says that this entire claim is false. It takes in all of the other claim and makes an additional one on top of it. So, in the formula on the right the scope of the tilde is *wider than the scope of the quantifier and that makes it the main operator for that formula*. This gives us our definition of the main operator. **The main operator of a statement is the operator with the widest scope**.

And take the term 'wider' literally. You can actually measure and tell which operator is wider. You know that both *quantifiers and the tilde come directly in front of the statements they operate on*. So you can draw lines from the operators over the statements they operate on, and see which is wider. Taking the two statements we just used, we get out our tape measure and measure the scope.

$$(\exists x)\underline{\sim Lxb} \qquad \underline{\sim(\exists v)Lvb}$$

So, just as before, we *see* that for the WFF on the left, the existential quantifier has the widest scope. For the WFF on the right, the tilde has the widest scope.

If you follow the rules correctly, there will never be a case in which you have two or more operators with identical scope in a single formula, and so you will always have only one main operator. The main operator is important for a number of reasons, many of which will become clear in later chapters. One of the functions of a main operator is that it allows us to distinguish between *kinds* of statements. A statement whose main operator is:

The tilde, is a **negation**.
The existential quantifier, is an **existential statement**.
The universal quantifier, is a **universal statement**.

So, after you have determined the RD, identified all the individuals, and created predicate functions for all the predicate relations, you then can turn to the analysis of the statements. Look to see if there are any operators. If there are, then the same principles will apply whether we have just a tilde, or a quantifier or some combination of universal quantifiers and tildes or universal and existential quantifiers and tildes. What is critical is ascertaining the main operator of the statement and making sure that this is reflected in the way in which you symbolize the proposition. Finally, remember that operators operate over *statements*, not *parts* of statements. You won't ever have a tilde just attached to a predicate or a name by itself.

## Exercise 4c

List the main operator for the following propositions and then identify what kind of a proposition it is, (e.g., negation, universal statement, existential statement).

1. ~(x)Pxy
2. ~(∃z)~Mzc
3. (∃y)(x)Mxy
4. (∃w)~Pwb
5. ~~(x)~Pxm
6. (x)~(∃y)Mxy
7. (x)(y)(z)~Cxyz
8. (∃w)~(x)(y)~Cwxy
9. ~(v)(w)~~(x)Cvwx
10. (x)Pxa

### EXTRA PRACTICE
Go back to Chapter Three and identify the main operators in all the exercises.

### ALTERNATE TRANSLATIONS

There's a further complication: sometimes there can be more than one acceptable symbolization of a statement. As an example, consider the following sentence:

Nobody's perfect.
~(∃x)Px
(x)~Px

1. Bob doesn't like anybody.

In sentence 1, you can easily see that there is both a negation and a universal statement, so you'll need to decide which one is going to be the main operator. Begin by asking yourself whether it is saying that it's false that Bob likes everyone or that for any person you find, it will be the case that Bob doesn't like them. If you think it's the former, then the main operator will be the tilde, and if you think it's the latter then the main operator will be the universal quantifier.

The correct answer is the second version, and so it would look like this:

**1. (x)~Lbx**

which in "logicese" would read, "For any person x, it's not the case that Bob likes them."

If we had the tilde as the main operator it would look like this:

**~(x)Lbx**

This reads, "It's false that for any person x, Bob likes them." In less stilted language, this says, "It's not true that Bob likes everyone." Of course even if Bob doesn't like everyone, it wouldn't rule out that there are lots of people that he does like. Hence, this *would not* be an accurate translation of sentence 1.

However, even though that would not be a correct translation, it turns out that there is another equally correct symbolization of sentence 1. What is another way of saying that Bob dislikes everybody? Another way to express this is by saying that not one single person exists who Bob likes or to put it still another way, "*It is false that there exists some person that Bob likes.*" Now, instead of a universal quantifier and a tilde, you have an existential quantifier and a tilde. How would this be symbolized?

It is false that: ~
there exists some person: (∃x)
that Bob likes: Lbx

Putting it together we have:

**~(∃x)Lbx**

This would be as acceptable a symbolization of sentence 1 as was our earlier one,

**(x)~Lbx**

Now, consider this sentence:

2. Not everybody likes Bob.

Again there are a couple of ways to express this statement in LOLA. The first is as a negation: It is false that everybody likes Bob.

$$\text{It is false: } \sim$$
$$\text{that everybody: } (x)$$
$$\text{likes Bob: Lxb}$$

$$\sim(x)\mathbf{Lxb}$$

The second way is as an existential statement: Someone exists who it's false that they like Bob, or simply, Somebody doesn't like Bob.

$$\text{Somebody: } (\exists x)$$
$$\text{Doesn't like Bob: } \sim Lxb$$

$$(\exists x)\sim\mathbf{Lxb}$$

In general, then, there are the LOLA variants for statements of this kind:

| No one | Not everyone |
|---|---|
| $\sim(\exists x)$ | $\sim(x)$ |
| $(x)\sim$ | $(\exists x)\sim$ |

As we continue to add to our LOLA vocabulary, we will have not only a wider range of statements we can express in LOLA, but also greater flexibility in expressing those statements. As we said in Chapter One, a statement can be expressed in many different natural language sentences; in LOLA too we will often have multiple LOLA WFFs that can adequately capture a statement. This gives us more freedom of expression in LOLA, but as the statements we translate get more complex, and we have more options, it often means we also need more ingenuity to translate statements correctly.

## Exercise 4d

Using this translation scheme, symbolize the following sentences.

**TRANSLATION SCHEME**
RD: Everything

| s: Santa Claus | p: Paul Bunyan | e: the Easter Bunny | a: Paul's axe |
|---|---|---|---|
| Px: x is popular | Mxy: x is more popular than y | Fxy: x has more fun than y | Gxyz: x gave y to z |

1. Nothing is popular.
2. Everything is popular.
3. Some things are not popular.
4. Santa Claus is not popular.
5. Santa Claus is more popular than the Easter Bunny.
6. Nothing is more popular than Santa Claus.
7. The Easter Bunny has more fun than Paul Bunyan.
8. Paul Bunyan doesn't have more fun than Santa Claus.
9. Nothing has more fun than Santa Claus.
10. Paul Bunyan gave his axe to the Easter Bunny.
11. Santa Claus didn't give anything to the Easter Bunny.
12. It's not true that Santa Claus didn't give something to the Easter Bunny.
13. Something gave Paul Bunyan's axe to the Easter Bunny.
14. It's false that everything is unpopular.
15. Everything is unpopular.

## Exercise 4e

Use the following translation scheme to translate the formulas below into English language sentences.

**TRANSLATION SCHEME**
RD: People

| j: John | s: Susan | p: Patty | Gx: x is a gossip |
|---|---|---|---|
| Lxy: x is a friend of y | Nxy: x is nicer than y | Txyz: x is telling the truth about y when talking to z | |

1. Gj
2. ~Gs
3. (∃y)~Gy
4. ~(x)Gx
5. Lsj
6. ~Ljs
7. (x)(∃y)Nxy
8. ~(∃z)(y)Nzy
9. ~Tpjs
10. (∃w)~Twjs
11. (x)(∃y)(∃z)Txyz
12. ~(x)(∃y)(∃z)Txyz
13. (y)~Tjsy
14. ~(∃z)Gz
15. (∃x)(∃y)~Lxy

# Exercise 4f

Construct a translation scheme for the following sentences and then symbolize them.

1. Some birds have wings.
2. All birds have wings.
3. Not all birds have wings.
4. The penguin at the zoo does not have wings.
5. Some birds do not have wings.
6. The penguin at the zoo is not faster than some birds.
7. All birds are faster than some bird.
8. No bird is faster than all birds.
9. Not every bird is faster than all birds.
10. It's not true that the penguin at the zoo is not faster than any bird.

> **THINKING AHEAD**
>
> What would you do with a sentence that has an "and" in the middle joining two different propositions? Can you do it with what we have in LOLA thus far? What would be the difference between a sentence like "Susan is a tall mathematician" and "Some people are tall and some people are mathematicians"?

## CONJUNCTIONS

We conclude this chapter by introducing the ***conjunction*** whose symbol is the "**&**" (ampersand) sign. Conjunction in LOLA roughly corresponds to such English language terms as: *and, but, yet, as well, also, in addition*. What is more, like English language conjunctions, the LOLA *conjunction joins two propositions together into a single statement*. However, there are important differences between it and our usual English language conjunctions. To begin with in English "and" can join sentence fragments, in LOLA conjunctions will only be *used to connect two complete statements*. Second, there are terms that are sometimes listed as conjunctions in English grammar that aren't considered conjunctions in LOLA, like *nor* and *or*. Third, certain grammatical signs, like the comma or semi-colon, while not defined as conjunctions in English grammar will be symbolized with an **&** in LOLA. The main point about conjunctions in LOLA is that they are used to join two propositions together.

> Bob lied and someone found him out.
> Lb & (∃x)Fxb

The ampersand, just like quantifiers and the tilde, has a scope. Unlike the quantifiers and the tilde, which operate over one statement that comes to their immediate right, the *scope of the ampersand is the two statements on either side of it*. In the

WFF below that expresses the statement, "Nothing's perfect, yet Paris is magnificent" (RD: Everything   Px: x is perfect   Mx: x is magnificent   p: Paris) we can see that the ampersand has the widest scope.

$$(x){\sim}Px \ \& \ Mp$$

Just as you could have more than one tilde in a formula, so also with the conjunction. For example, you can have two propositions that are joined by an ampersand to form a statement that is itself then joined to another statement by a conjunction to form a new proposition, which is then joined to yet another statement by a conjunction and so on. Furthermore, as with quantifiers and the tilde, the conjunction will sometimes be the main operator and sometimes it will be a part of a larger proposition, such as a negation.

---

**Example Set 4b**
TRANSLATION SCHEME
RD: Everything

| a: Allen | Px: x is a person | Mx: x has mental qualities | Cx: x is made of corporeal substance |
|---|---|---|---|

1. Some things are made of corporeal substance and some things have mental qualities.
2. Some things are made of corporeal substance, but also have mental qualities.
3. Some people have mental qualities.
4. Everything is both mental and corporeal.
5. Allen is made of corporeal substance, yet doesn't have any mental qualities.
6. Some people have mental qualities and are made of corporeal substance.

---

If you have already been given a translation scheme, then the first task is to identify the operators in the sentence. Do you have any quantifiers, any tildes, or any conjunctions? In sentence 1, there are two existential quantifiers, no tildes, and a conjunction. The next task is to determine the *main* operator. Sentence 1 is an assertion of *two* separate propositions that are then joined together via a conjunction. This makes the conjunction the main operator. When a conjunction is the main operator, the best strategy is to symbolize the two propositions one at a time. We'll start with the statement, "Some things are made of corporeal substance":

$$(\exists x)Cx$$

The second is, "Some things have mental qualities," which would be symbolized as:

$$(\exists y)My$$

Now, place the & sign between the two formulas and you have the complete WFF:

$$(\exists x)Cx \,\&\, (\exists y)My$$

Sentence 2 seems similar, but the main operator is not the conjunction. The claim this time is that there exists a thing that is *both* made of corporeal substance and possesses mental qualities. So, the existential quantifier is the main operator. You can symbolize sentences in whatever way is most effective for you, but one way to symbolize such a statement is to begin by putting down your existential quantifier,

$$(\exists x)$$

Having said that there exists some x, you then have to symbolize the claims that are to be made about this x. You want to assert that this x has two sets of properties, and at the moment you don't have a technique for asserting two distinct sets of properties about something.

> Parentheses enclose two statements and indicate scope.

What we will do is return to the notion of the scope of an operator, and *add a way to indicate the extent of an operator's scope*. Symbolize the claim that x is made of corporeal substance and x has mental qualities and place these next to your existential quantifier:

$$(\exists x) \; Cx \,\&\, Mx$$

Now, place one parenthesis to the left of the Cx and one parenthesis to the right of the Mx, thereby giving you:

$$(\exists x)(Cx \,\&\, Mx)$$

What the parentheses do is to say that the *scope of the operator to the immediate left of them extends over whatever formula is located within them*. If you left out the parenthesis, as in $(\exists x)Cx \,\&\, Mx$, the existential quantifier would only range over the sentence immediately following it: Cx. The Mx is left "hanging"; it *isn't bound to the quantifier* and, therefore, Mx is not a WFF; because the & does not connect two WFFs, it is not well-formed either.

We'll show this pattern again by symbolizing sentence 3. Sentence 3 claims that something exists that is *both* a person and has mental qualities. Once again, you write down your existential quantifier,

$$(\exists z)$$

and then you symbolize what it is you want to claim about this thing that exists, namely it is a person *and* it has mental qualities:

$$(\exists z) \; Pz \; \& \; Mz$$

Then put in the parentheses to show the scope of the quantifier:

$$(\exists z)(Pz \; \& \; Mz)$$

Sentence 4 is pretty much the same, although this time it uses a universal quantifier, thereby generating,

$$(y)(Cy \; \& \; My)$$

**Remember:** *the parentheses are not optional.* Universal and existential quantifiers operate over the statements that *directly follow them*. If we wrote sentence 4 as (y)Cy & My, the statement directly following (y) is only Cy, because the & indicates "another statement coming up." And My would be a separate statement. This would be wrong for two reasons: (1) My, without a quantifier, is not a WFF. The variable, y, must be bound to a quantifier (which in turn refers back to an RD) in order to have a meaning and be a complete statement. The parentheses assure that the My is bound to the (y). And (2) since it would be unbound and doesn't refer back to the (y) it of course doesn't adequately represent the statement expressed in sentence 4.

Sentence 5 raises new issues as it doesn't reference any quantifier. Proceed as you normally would by identifying the operators and then determining which one is the main operator. There is a conjunction and a tilde, which one of these has the largest scope? The conjunction is connecting both the statements and the tilde is only operating over the statement about Allen's mental qualities, hence the conjunction is the main operator. So, symbolize both statements individually:

$$Ca \qquad \sim Ma$$

Now join the two together with the conjunction, resulting in:

$$Ca \; \& \; \sim Ma$$

Does this require parentheses around it? No, you have already unambiguously captured the original sentence, which is an assertion that Allen possesses two different traits. It is clear that the main operator of this statement is the ampersand and that the statement is a conjunction.

Finally, sentence 6 uses a quantifier and it assigns three attributes to the thing. Strictly speaking,

$$(\exists v)[Pv \ \& \ Cv \ \& \ Mv]$$

is not well formed (and neither is A & B & C). An extra set of parentheses shows exactly what statements are connected by what ampersands, so we should write

$$(\exists v)[Pv \ \& \ (Cv \ \& \ Mv)]$$

We could just as well have divided things differently:

$$(\exists v)[(Pv \ \& \ Cv) \ \& \ Mv]$$

Either way expresses the same thing, in this case.

An additional advantage of the conjunction is that it frees us up from some restrictions regarding our translation schemes. Prior to this innovation, we were not able to adequately translate a "mixed" discourse of, say, things, people, and cities. Now, within certain limitations, we will be able to have sets of sentences with a wider variety of references within them.

## Exercise 4g

Use the following translation scheme to symbolize sentences 1-10.

**TRANSLATION SCHEME**
RD: Everything

| h: Hannibal | t: Tom | f: Fred |
|---|---|---|
| Ex: x is an electrical device | Px: x is a person | Dx: x is dangerous |
| Sx: x is smart | | |

1. Hannibal is dangerous and smart.
2. It's not true that everyone is dangerous as well as smart.
3. Some electrical devices are dangerous.
4. Some people are not dangerous.
5. Some people are dangerous but not smart.
6. It's not true that all people are dangerous.

7. Fred is smart, although he's not dangerous.
8. Tom is someone who is not only not dangerous, he's not smart.
9. It's simply untrue that some people are not dangerous.
10. Some things are dangerous but not smart.

### THINKING AHEAD

Science often expresses its conclusions in an if ... then format; for example, "If a planet has a certain mass, then it will have the following gravity." How would you represent this kind of relationship and what kind of properties do you think it should have?

# Chapter Five: Conditionals and Disjunctions

**What's Up?**
If, then →
Or ∨

In this chapter we'll introduce two more operators of LOLA. Both operators function in a similar way to the conjunction, in that they join or connect two statements into one compound statement. Like conjunctions, these two operators

1. connect WFFs (*not* sentence fragments)
2. their scope is the WFFs on either side of the operator

## CONDITIONALS

Conditionals are what we term the "*if ... then*" relationship between statements, such as "If Kevin likes vacationing in warm places, then he should go to Cancun." Example Set 5a illustrates some of the many English language variations of the conditional operator.

### TRANSLATION SCHEME
RD: Everything

| b: Bob | a: Allen | k: Kevin |
|---|---|---|
| Px: x goes to the party | Ex: x gets an 85 on the final | Cx: x passes the course |
| Gx: x pays for the gas | Dxy: x drives y to college | Lxy: x gives y a lift |

**Example Set 5a**
1. Bob will go to the party, if Allen gives him a lift.
2. Allen giving Bob a lift is a sufficient condition for Bob going to the party.
3. Getting an 85 on the final is a necessary condition for Allen to pass the course.
4. Assuming Allen passes the course, he's got an 85 on the final.
5. Allen will drive Kevin back to college on the condition that Kevin pays for the gas.
6. Kevin will pay for the gas only if Allen drives him back to college.
7. If Allen goes to the party, then he passed the course.
8. Allen won't go to the party unless he passes the course.

In each case you have two statements that have a special kind of relationship to each other. All conditionals express, in one way or another, the idea that *if one condition is or comes to be, then a second condition also is or comes to be*. The statement following the 'if' is called the **antecedent** of the conditional, and the statement following the 'then' is called the **consequent** of the conditional. It is not necessarily the case that the antecedent will occur first in an English declarative sentence; it may well be that the consequent is written first. This is the case in sentences 1, 3, and 5. The important thing to ascertain is the nature of the relationship between the statements. Ask yourself, which statement is the dependent one, which one requires the other one to happen or be before it can take place? The symbol used to represent the conditional relationship is →, called "the arrow," and it goes after the antecedent and before the consequent:

> If you don't bet, you can't lose.
> (x)(~Bx → ~Lx)

**Antecedent → Consequent**

This results in the following translations of the sentences in Example Set 5a:

1 & 2: Lab → Pb
3 & 4: Ca → Ea
5 & 6: Gk → Dak
7 & 8: Pa → Ca

(**Study these examples carefully**. It is a very good idea to become familiar with the different English variants that → expresses and which statement is the antecedent and which the consequent in each variant.)

The terms "sufficient" and "necessary" are so commonly used that we need to be clear about how they are to be symbolized. If something is designated a *sufficient condition*; it is *enough* to ensure another condition comes to be, so it is the *antecedent* of the conditional. When a statement is called a *necessary condition*, it is *needed* for another condition to be, so it is the *consequent* of the conditional.

Sometimes it helps to keep in mind some obvious examples to keep the order straight. For example, think of something that makes you happy (say, like having chocolate ice cream) no matter what else is going on. That thing is *sufficient*, is *enough*, for happiness, and would be the antecedent of the conditional:

**Ia → Ha**

For necessary conditions, think about fire. You *need* oxygen to build a fire, so it's a *necessary* condition, and would be the *consequent* of the conditional:

## CHAPTER FIVE: CONDITIONALS AND DISJUNCTIONS

$$Fa \rightarrow Oa$$

(If this still doesn't sound quite right, think if we translated this into a conditional that was written the other way: Om → Fm. That means if there is oxygen around, there will be fire, which would make breathing quite hazardous!!)

Also, one of the expressions above, "only if" is a bit unusual. While "if" usually indicates that the statement following it is the antecedent, "only if" has a different connotation. The statement following "only if" indicates that it is the necessary condition of the conditional and consequently it's designated as the consequent of the conditional. Think of "only if" as a single unit, with an arrow written over it, indicating to read left to right, antecedent to consequent:

**Antecedent** $\overrightarrow{\text{only if}}$ **Consequent**

Finally "not ... unless" statements can be expressed as the conditional relationship because they are asserting that condition X *won't* obtain *unless* condition Y does; or, in other words, ***if*** condition X does obtain, ***then*** we know that condition Y does also.

The conditional, like the conjunction, is also frequently used together with quantifiers. Consider the following statements,

### TRANSLATION SCHEME
RD: Everything

| Tx: x is a tree | Px: x is a plant | Hx: x is a person | Dx: x is a dog | Wx: x will have its day |
|---|---|---|---|---|
| Fx: x wants a fulfilling career | Ax: x is alive | Rx: x has rights | Gx: x is good | Nx: x wins |

1. All trees are plants.
2. People want to find a fulfilling career.
3. Every dog will have its day.

All of these can be expressed using conditionals since they are all of the same basic structure, *if* x is this, *then* it's also this. However, the main operator is not the conditional, but the universal quantifier, which means you'll need to place parentheses in the correct spot.

This gives the following symbolizations,

1. **(x)(Tx → Px)**
2. **(x)(Hx → Fx)**
3. **(x)(Dx → Wx)**

Which in "logic speak" are,

1. For any x you find, if x is a tree, then it's a plant.
2. For any x you find, if x is a person, then x wants a fulfilling career.
3. For any x you find, if x is a dog, then x will have its day.

Often students ask, why the arrow and why not the ampersand? After all, aren't we saying here that any tree is *both* a tree and a plant? And that any person is *both* a person and wants a fulfilling career? And that any dog is *both* a dog and will have its day? But see what we get if we translated these statements with the ampersand:

1. (x)(Tx & Px)
2. (x)(Hx & Fx)
3. (x)(Dx & Wx)

We now know, however, that if these were our translations, they would read:

1. *Everything* is both a tree and a plant.
2. *Everything* is a person with a fulfilling career.
3. *Everything* is a dog that's having its day.

These, of course, do not correctly capture the original statements.

Sometimes, however, the ampersand *is* used with universal statements, *particularly when the RD is restricted*. For example, if you had the RD: People, and you wanted to translate the statement, "Everybody is good and has rights," you would symbolize this:

(x)(Gx & Rx)

But when you're tempted to translate a statement into LOLA by a universal statement with an ampersand in its scope, you should examine it carefully to make sure it expresses what you meant.

Returning to our translation scheme with RD: Everything, consider these next two statements:

4. If everything is alive, then something has rights.
5. If something is good, then everything wins.

These propositions, though there are quantifiers in them, are conditionals. The arrow is the main operator in these statements, the operator with the widest scope. So, they would be symbolized:

**4. (x)Ax → (∃y)Ry**
**5. (∃y)Gy → (z)Ny**

Finally, using the tilde, the arrow, and the universal quantifier allows you to symbolize universal negative claims such as:

6. No dog is a plant.

This is another way of saying that *if* something is a dog, *then* it is not a plant. This is symbolized as,

**6. (x)(Dx → ~Px)**

Note, by the way, that there's another equally correct way of translating this. We can rephrase the English sentence into 'No dog is a plant', and into logic-speak as 'It's not the case that there exists something that's both a dog and a plant, so into LOLA as ~(∃y)(Dy & Py).

# Exercise 5a

Use the following translation scheme to symbolize sentences 1-10.

**TRANSLATION SCHEME**
RD: People

| a: Allen | k: Kathy | Ty: y likes tea |
| Cz: z likes coffee | Hw: w is happy | Jx: x is jittery |

1. Everyone who likes coffee is jittery.
2. Kathy doesn't like tea.
3. Some people will be happy only if everyone is unhappy.
4. Allen is happy only if everyone is happy.
5. No one who likes tea likes coffee.
6. Kathy will be happy if Allen is happy.
7. Somebody likes coffee and is not jittery.
8. Everyone likes coffee if Allen does.
9. Not everybody who likes coffee dislikes tea.
10. If Allen likes tea, then he's happy.

Using the same translation scheme as above, translate the following WFFs into English.

11. (∃z)Tz → (∃x)~Cx
12. (x)(Tx → Hx)
13. (y)Ty
14. (z)(Cz → ~Tz)
15. ~(v)(Cv → ~Hv)
16. Ck → Jk
17. (x)Cx → Ca
18. (v)Tv → ~(x)Cx
19. Ta & Tk
20. (x)Tx → (∃z)(~Cz & ~Jz)

## Exercise 5b

Construct a translation scheme for the following sentences, and then symbolize them.

1. Everything that is alive needs oxygen.
2. All living things must reproduce.
3. Some things reproduce, but don't mate.
4. If a thing mates, it must be alive.
5. Rover is a living thing.
6. Some things are alive.
7. Assuming that Rover reproduced, she mated.
8. If some things mate, then at least one thing is alive.
9. Whatever needs oxygen is a living thing.
10. There are no living things that don't reproduce.
11. Not everything that needs oxygen is alive.
12. It's not the case that Rover is alive and doesn't need oxygen.
13. Rover is a living thing and she reproduced.
14. Nothing that reproduces is not alive.
15. A sufficient condition for needing oxygen is being alive.
16. A thing has reproduced only if it mates.
17. A necessary condition for being alive is reproduction.
18. Everything is alive if everything needs oxygen.
19. Rover is not alive.
20. If Rover is not alive, then something is not alive.

### THINKING AHEAD

What would you do with a sentence that used an "or" instead of an "and"? Can you use the same symbol? What are the differences between "or" and conjunctions? What would you do with a "neither ... nor" sentence? Would you need a negation as well as another symbol?

## DISJUNCTIONS

Disjunctions are the types of statements that in English are often expressed by the English word 'or.' The LOLA symbol for the disjunction is the wedge, **v**.

> If Bob doesn't get the job, then either Mary or Lionel will.
> ~Jb → (Jm v Jl)

What makes 'or' statements a bit tricky is that, in English, we switch between two senses of the word 'or' all the time without much conscious thought. These two different ways of using "or" are represented in sentences 1 and 2.

1. Either the Detroit Lions or the Philadelphia Eagles will win the Super Bowl.
2. He must be crazy or stupid.

Sentence 1 expresses what we call *the exclusive* sense of 'or.' In statements like these, part of the meaning of the proposition is that *only* one of the two statements will be the case; either the Lions will win or the Eagles will win, *but not both*. The second sentence is the *inclusive* sense of 'or', and the idea behind this one is that one, or the other, or *both* are the case. In our logical system, the **v** represents the *inclusive* sense of "or." The exclusive version is easily handled by a combination of a disjunction, tilde, and conjunction, as we'll see below.

Unlike the conditional or the conjunction, there are not that many English language variations of "or." Basically there are just two: "or" and "unless". For example,

1. You need four years' experience or a college degree.
2. You need four years' experience unless you have a college degree.
   (Ex: x has four years' experience; Cx: x has a college degree)

These two statements make essentially the same claim, and hence they are both symbolized using the wedge, giving you,

**(x)(Ex v Cx)**

As always, it is critical to ascertain the main operator of the statement in order to render a successful symbolization of the sentence. For example, consider the following three sentences and ask yourself, what is the main operator?

RD: Everything

3. Kansas will secede from the Union unless Lincoln can persuade the governor it is a bad idea. (Pxy: x persuades y that secession is a bad idea; Sx: x will secede from the Union; k: Kansas; l: Lincoln; g: governor of Kansas)

4. A mammal has a kidney or a heart. (Mx: x is a mammal; Kx: x has a kidney; Hx: x has a heart)
5. Either everything is physical or something is spiritual. (Px: x is physical; Sx: x is spiritual)

Sentence 3 is a disjunction, the main operator of the statement is the claim that either one of the statements or both of them will happen. Hence, it is symbolized as,

### 3. Sk v Plg

Sentence 4 is making a general claim about all mammals and so its main operator is the universal quantifier. It's saying that for anything x that happens to be a mammal, it will have one or the other or both of these traits. So, its main operator is a universal, the operator with the next largest scope would be the arrow, and then finally the wedge. This gives you,

### 4. (x)[Mx → (Kx v Hx)]

Sentence 5 also combines quantifiers with wedges, but in this case the wedge is the main operator. As is the case with conjunctions and conditionals, try to symbolize the two component statements and then join them with the wedge. The first of the statements is "everything is physical" which would be

### (x)Px

and the remaining part is "Something is spiritual" which would be,

### (∃x)Sx

Now, simply place the wedge between them and you're done,

### 5. (x)Px v (∃x)Sx

There are two commonly used English phrases that involve "or" that need to be addressed. The first is the exclusive sense of "or" and the second is the phrase, "neither nor". What is interesting about these two locutions, is that neither of them has a **v** as its main operator. Consider the following example,

6. Either Sam will be in the library or he'll be at the concert at 9 PM. (s: Sam; Lx: x is in the library at 9PM; Cx: x is at the concert at 9PM)

Given that it's impossible for a person to be in two places at the same time, it's reasonable to say that the intent of this sentence is the exclusive sense of "or". However, since our default symbolization of "or" is the inclusive sense, we need to make *it explicit if we mean the exclusive sense* in a case like sentence 6. If you were to write out what the exclusive "or" means it would be, "Either option 1 or option 2 is the case, but *not both* of them." Sometimes the "*but not both* of them" will be stated explicitly; other times it won't be stated, but you'll be told to symbolize the sentence using the exclusive sense. Interpreting sentence 6 as the exclusive sense of "or" makes the first statement, "Either Sam will be in the library *or* he'll be at the concert at 9PM" and the second, "*It is not true that* Sam will *both* be in the library *and* that he will be at the concert at 9PM." *Notice that exclusive sense makes the conjunction the main operator, not the disjunction.* So translate both sides of the sentence and then place a conjunction in between them.

> Exclusive 'or'
> (_ v _) & ~(_ & _)

This gives you,

**Ls v Cs**

on one side.

Now you have to translate the claim that "It's not the case that Sam is in the library and that Sam is at the concert" which has both a conjunction and a tilde as part of it. The conjunction has the smaller scope, so symbolize it first

**Ls & Cs**

Now, state that this claim is false. To do that you have to put parentheses around the statement and put a tilde to the left of it, rendering the following

**~(Ls & Cs)**

This gives you both sides of the statement. Since the conjunction will be the main operator, you need to make sure that its scope extends over both sides of the statement. This means you'll need to put parentheses around Ls v Cs rendering,

**(Ls v Cs)**

You can now bring the two sides together with a conjunction between them, thereby completing the symbolization

**6. (Ls v Cs) & ~(Ls & Cs)**

"Neither nor" symbolizations can be done one of two ways. Consider the following example,

7. Neither France nor England has the largest economy in the world.
(f: France; e: England; Lx: x has the largest economy in the world)

> Neither ... nor
> ~(_ v _)
> ~_ & ~_

This statement says that it is false that France has the largest economy in the world *and* it is also false that England has the largest economy in the world. This means you have a conjunction as your main operator, which is joining two negated statements. Simply symbolize the two negated statements

$$\sim Lf$$
$$\sim Le$$

and then join them with a conjunction and you have,

**7. ~Lf & ~Le**

It is also equally acceptable to say that the above statement is a way of saying, "It's false that either France has the largest economy in the world or that England has the largest economy in the world." In this case, you have a negation of a disjunction—a "not or" as it were. So, in this case you'd symbolize the disjunction first,

**Lf v Le**

and then assert that the above claim is false.

**7. ~(Lf v Le)**

> Not ... unless
> ~_ v _
> _ → _

Finally, as you saw in Example Set 5a, number 8, the section on conditionals, "not unless" can be symbolized using the arrow. But "unless" can also be symbolized as a variation of "or," so "not unless" statements can be symbolized using the tilde and the wedge. So for the statement,

8. The cat won't go out unless the dog is asleep. (c: the cat, d: the dog, Gx: x goes out, Ax: x is asleep)

we can symbolize it in either of the two following ways:

**8. ~Gc v Ad**
**8. Gc → Ad**

# CHAPTER FIVE: CONDITIONALS AND DISJUNCTIONS

> **THINKING AHEAD**
>
> How would you represent the idea that Kathy is the same size as herself, or that if 1<2, and 2<3, then 1<3 or that if Jim is Fred's dad, then Fred can't be Jim's dad. How would you represent these general kinds of relationships in LOLA?

## REFLEXIVITY, SYMMETRY, AND TRANSITIVITY

Before moving onto the last two operators of LOLA, we'll examine three types of logical relations, which we can now symbolize with the addition of the conditional and disjunction.

---

**Example Set 5b**
TRANSLATION SCHEME
RD: People

| Sxy: x is the same size as y | Nxy: x is near to y | Oxy: x is older than y |
|---|---|---|

| Bxy: x is a sibling of y | Axy: x is the same age as y | |
|---|---|---|

1. Everyone is the same age as him or herself.
2. "Being the same age as" is reflexive.
3. If one person is near a second person, then the second person is near the first person.
4. "Being a sibling of" is a symmetric relationship
5. If one person is older than a second, and that second person is older than a third, then the first person is also older than the third.
6. The relation "being the same size" is transitive.

---

Sentences 1 and 2 say the same thing, and are examples of the reflexive relation. Reflexivity is a *property* of certain relations. Whenever something stands in any sort of relationship to itself, that relation is a reflexive one. Sentences 1 and 2 could be symbolized,

### 1. and 2. (x)Axx

Notice that you *use the same variable twice* to indicate that this relationship is bearing it to itself.

Sentences 3 and 4 are examples of symmetric relations. Symmetry is a *property* of certain relations. A symmetric relationship is one in which you have two things and they both bear the same relation to each other. What you're saying, in essence, is that **if** any y has a certain relationship to any z, **then** z has that same relationship to y. We use two different universal quantifiers and the arrow to express these relations:

3. (y)(z)(Nyz → Nzy)
4. (x)(w)(Bxw → Bwx)

Sentences 5 and 6 are examples of transitive relations. Transitivity is a *property* of certain relations. The transitive relation involves comparing three entities. We express transitive relations in LOLA using three universal quantifiers, and both the arrow and the ampersand:

5. (x)(y)(z)[(Oxy & Oyz) → Oxz]
6. (u)(v)(w)[(Suv & Svw) → Suw]

There are, of course, many relations that are not reflexive, others are not symmetric, still others are not transitive, and we'd like to be able to say something, in LOLA, about these kinds of relations. We'll show some of these relations using the following Example Set.

### Example Set 5c
**TRANSLATION SCHEME**
RD: People

| Gxy: x is greater than y | Fxy: x is faster than y | Pxy: x is the father of y |
|---|---|---|

1. Nobody can be greater than him or herself.
2. If one person is faster than a second, the second can't be faster than the first.
3. If one person is the father of a second, and a second is the father of a third, then the first can't be the father of the third.

From what we've learned about the reflexive, symmetric, and transitive relations, we can say about the relations in our three sentences that:

1. "Being greater than" is *not* a reflexive relation.
2. "Being faster than" is *not* a symmetric relation.
3. "Being the father of" is *not* transitive relation.

And we can symbolize them in LOLA as:

    1. (x)~Gxx
    2. (x)(y)(Fxy → ~Fyx)
    3. (x)(y)(z)[(Pxy & Pyz) → ~Pxz]

And we'll designate relations of these types as

1. Irreflexive
2. Asymmetric
3. Intransitive

> **Example Set 5d**
> TRANSLATION SCHEME
> RD: People
>
> | Lxy: x loves y | Fxy: x does a favor for y | Exy: x is the enemy of y |
> |---|---|---|

There are some relations where we cannot say that they are reflexive or not, symmetric or not, or transitive or not. Take the relation "loves." Sometimes people love themselves, sometimes they don't. Sometimes when one person loves another, that other person loves them, but sometimes they don't. Sometimes when one person loves a second, and that second person loves a third, the first will also love the third, but sometimes they don't. We can state the indeterminate quality of the relations, as in the following:

1. People may or may not love themselves.
2. If a person does a favor for another, that other may or may not return the favor.
3. If one person is the enemy of a second, and the second is the enemy of a third, the first may or may not be the enemy of the third.

And we can translate them to LOLA as:

    1. (x)(Lxx v ~Lxx)
    2. (x)(y)[Fxy → (Fyx v ~Fyx)]
    3. (x)(y)(z)[(Exy & Eyz) → (Exz v ~Exz)]

We'll designate relations like these as non-reflexive, non-symmetric, and non-transitive, respectively.

> **Example Set 5e**
> For the following example sentences, try translating them into LOLA before looking at the answers.
> **TRANSLATION SCHEME**
> RD: Everything
>
> | l: Lucy | j: Jim | b: Bob | r: Russia |
> |---|---|---|---|
> | a: United States | i: Iran | Lxy: x likes y | Rxy: x respects y |
> | Hxy: x has heard of y | Sxy: x is a sister of y | Exy: x supports y | Fxy: x is friends with y |
>
> 1. Lucy may or may not respect herself.
> 2. The "liking" relation is a non-reflexive relation.
> 3. If Bob has heard of Jim, Jim may or may not have heard of Bob.
> 4. "Being a sister of" is a non-symmetrical relationship.
> 5. If the US supports Russia, and Russia supports Iran, the US may or may not support Iran.
> 6. The relationship of "being friends with" is non-transitive.
> 7. Jim does not respect himself.
> 8. "Respects" is a non-transitive relationship.
> 9. "Being friends with" is a non-symmetrical relationship.
> 10. Russia doesn't respect the United States.

We'd represent these in LOLA as:

1. **Rll v ~Rll**
2. **(x)(Lxx v ~Lxx)**
3. **Hbj → (Hjb v ~Hjb)**
4. **(x)(y)[Sxy → (Syx v ~Syx)]**
5. **(Ear & Eri) → (Eai v ~Eai)**
6. **(x)(y)(z)[(Fxy & Fyz) → (Fxz v ~Fxz)]**
7. **~Rjj**
8. **(x)(y)(z)[(Rxy & Ryz) → (Rxz v ~Rxz)]**
9. **(x)(y)[Fxy → (Fyx v ~Fyx)]**
10. **~Rra**

Look carefully at these examples to make sure you understand why each statement is symbolized as it is. If you are able to recognize and quickly translate these complex statements, you are well on your way to becoming fluent in LOLA!

In the next chapter we introduce a new operator that represents an important logical 2-place predicate relation that is reflexive, symmetric, and transitive. We'll also see how that new operator and comparative 2-place predicates can be used to express statements about "best in class" individuals.

The following exercises focus primarily on conditionals and disjunctions, though they will still include sentences that don't require either arrows or wedges.

## Exercise 5c

Using the following translation scheme, translate each sentence from English into LOLA.

**TRANSLATION SCHEME**
RD: Everything

| t: Tolstoy | d: Dostoevsky | c: Camus | p: *Crime and Punishment* | a: *War and Peace* |
| --- | --- | --- | --- | --- |
| s: *The Stranger* | Px: x is a person | Tx: x is a book | Bx: x is boring | Rxy: x reads y |
| Ax: x is an author | Ix: x is interesting | Gx: x will get his book back | Wxy: x wrote y | Lxyz: x loans y to z |

1. Either Tolstoy wrote *Crime and Punishment* or Dostoevsky did, but not both of them.
2. If someone boring wrote *War and Peace*, then it's a boring book.
3. If Tolstoy wrote *War and Peace*, then he didn't write *Crime and Punishment* or *The Stranger*.
4. It's not true that Dostoevsky wrote *The Stranger*, although Tolstoy did write *The Stranger*.
5. Camus wrote *The Stranger*, but not *War and Peace*.
6. If someone read *War and Peace*, then somebody read an interesting book.
7. *War and Peace* is an interesting book.
8. Someone is a boring author, but not Camus.
9. If anyone loans a book to Tolstoy, then that person will get his book back.
10. If anything is an interesting book, *Crime and Punishment* is.
11. Nothing interesting is boring.
12. Neither Camus nor Tolstoy wrote *Crime and Punishment*.
13. Every boring author is uninteresting.
14. Camus is a boring author, but interesting.

15. No boring author reads interesting books.
16. If anyone loans a boring book to someone, then they'll not get their book back.
17. If someone wrote *War and Peace*, then somebody wrote an interesting book.
18. If Tolstoy loaned his copy of *Crime and Punishment* to Camus, then somebody's not getting his book back.
19. No boring author writes interesting books.
20. A person who reads *The Stranger* has read an interesting book.

Using the same translation scheme, translate the following LOLA WFFs into English sentences.

21. Wts
22. (∃x)Wxs v ~(∃x)Wxs
23. (∃x)Rdx
24. (x)[(Ax & Ix) → (Ix & Px)]
25. Wdp v (∃x)[(Px & Ax) & Wxp]
26. (x)(∃y){[Px & (Ty & By)] → Rxy}
27. (x)[(Tx & Ix) → ~Bx]
28. (x)[Tx → (Bx v Ix)]
29. (x)[(Ix & Px) → {(Bx & Ax) v (~Bx v Ax)}]
30. Lcpt → ~(x)(Px & Rxp)
31. Tp & Ip
32. ~(∃x)Ax → ~(∃x)Wxa
33. (x)[Px → (∃y)(Ty & Rxy)]
34. (∃x)(~Tx & Wcx) v Wcs
35. (∃x){Px & (∃y)[Ty & (Lxyd v Lxyc)]}
36. (Rtc v Rdc) & ~(Rtc & Rdc)

## Exercise 5d

Using the following translation scheme, translate each sentence from English into LOLA.

**TRANSLATION SCHEME**
RD: People

| Ixy: x is identical to y | Mxy: x is the mother of y | Lxy: x loves y |

1. "Being identical to" is a reflexive relation.
2. "Being identical to" is a symmetrical relation.
3. "Being identical to" is a transitive relation.
4. "Being the mother of" is an irreflexive relation.
5. "Being the mother of" is an asymmetrical relation.
6. "Being the mother of" is an intransitive relation.
7. "Loving" is a nonreflexive relation.
8. "Loving" is a nonsymmetrical relation.
9. "Loving" is a nontransitive relation.

# Exercise 5e

Using the following translation scheme, translate each sentence from English into LOLA.

**TRANSLATION SCHEME**
RD: People

| a: Albert | b: Burt | c: Carmen | Hxy: x helps y |
| --- | --- | --- | --- |
| Pxy: x is the partner of y | Axy: x is the ancestor of y | Cx: x goes to college | |

1. "Helps" is a nonreflexive relation.
2. Albert and Burt are partners.
3. "Ancestor of" is an intransitive relation.
4. Carmen will help Albert only if Burt also helps Albert.
5. Burt won't go to college unless neither Albert nor Carmen goes.
6. Carmen will go to college if Albert or someone goes.
7. Someone is the ancestor of Carmen.
8. No one will go to college unless both Albert and Burt go.
9. If anyone goes to college, Carmen will go.
10. Burt's the ancestor of Carmen or the ancestor of Albert, but not both.

### THINKING AHEAD

Imagine combining the two phrases "if" and "only if". What symbol would you use to represent this relationship? What does it say? Think of some situations in which you would apply the phrase "if and only if".

# Chapter Six: Biconditionals and Identity

> **What's Up?**
> Biconditional ↔
> Identity =

The last two operators we'll introduce into LOLA are the double arrow (↔) and the identity symbol (=). These operators are each a bit different from those we've examined thus far.

## BICONDITIONALS

A biconditional expresses the relationship of one statement being so *if and only if* the other statement is so. Essentially then, the biconditional is a conjunction of two conditionals. For example the statement,

1. Bob can drive if and only if he pays for insurance.

could be rephrased as

2. Bob can drive **if** he pays for insurance, **and** Bob can drive **only if** he pays for insurance.

We could, of course, symbolize statement 2 using the → and the & (b: Bob; Dx: x drives; Px: x pays for insurance):

3. (Pb → Db) & (Db → Pb)

> This flu outbreak is a pandemic if and only if it's a global outbreak.
> Pf ↔ Gf

So, we don't strictly need to introduce the double arrow into LOLA in order to translate English biconditionals. There are good reasons, though, for introducing this new symbol. First, of course, it makes for shorter translations! More importantly, it highlights the relationship between the two statements—that the conditional relationship "works both ways," as it were. Further, it allows us to explicitly capture those English statements that express this two-way conditional relation. There are typically three linguistic expressions that convey this relationship.

### Example Set 6a
**TRANSLATION SCHEME**
RD: Everything

| s: Sam | d: Dave | f: Fred | Wy: y wins |
|---|---|---|---|
| Px: x is a person | Bx: x is a bearer of rights | Gx: x is guilty | Mx: x is the murderer |

1. Sam wins *if and only if* Fred doesn't.
2. Being a person is *a necessary and sufficient condition* for being a bearer of rights.
3. Dave is guilty *just in case* he's the murderer.

The biconditional is symbolized with the double arrow ↔ between two statements. As you examine propositions 1-3, remember to pay close attention to when the biconditional is the *main* operator and when some other operator is. For example, in sentence 1, the biconditional, not the tilde, is the main operator, so it results in:

**1. Ws ↔ ~Wf**

In sentence 2, the main operator is the universal quantifier and so it gives us,

**2. (x)(Px ↔ Bx)**

To see why this is so, symbolize it with the biconditional as the main operator which would give you the following,

**(x)Px ↔ (x)Bx**

This statement, when translated into English, would read, "Everything is a person if and only if everything is a bearer of rights," which is a different claim than the one asserted above.

Sentence 3 has the biconditional as its main operator and so it generates,

**3. Gd ↔ Md**

# Exercise 6a

Use the following translation scheme to symbolize sentences 1-20 into LOLA.

**TRANSLATION SCHEME**
RD: People

| m: Marvin | r: Ralph | j: Janice | Bx: x is a bachelor | Ax: x is an adult |
| --- | --- | --- | --- | --- |
| Mx: x is a male | Wx: x is married | Hx: x is happy | Hxy: x is happier than y | Wxy: x is married to y |

1. Ralph is not a bachelor, yet he's happier than Marvin who is.
2. A necessary and sufficient condition for being a bachelor is being an unmarried adult male.
3. Marvin is happier than Ralph if and only if Ralph is a bachelor and Marvin isn't.
4. Janice is married only if she's an adult.
5. It's untrue that "being happier than" is a symmetrical relation.
6. Ralph is a happy bachelor if and only if he is a happy, adult, unmarried male.
7. Janice can't be a bachelor if she's not a male.
8. Marvin is happy just in case he's married.
9. Janice is happier than everyone if and only if she's not married.
10. Married people are happier than unmarried people.
11. No one is happy if they are not married.
12. Not everyone is married if and only if somebody is unmarried.
13. No one is a bachelor just in case everyone is married.
14. Anybody who is not married, but not an adult, is not a bachelor.
15. Every single adult male is a bachelor, but even though Ralph is an unmarried male, he's not an adult.
16. If and only if Janice is a male is it the case that everyone is a bachelor.
17. Some bachelors are happier than someone who is married.
18. It's false that everyone is a bachelor if and only if they are unmarried males.
19. Janice and Ralph and Marvin are married adults, although Janice and Ralph are married to each other and neither one is married to Marvin.
20. No one is married to Marvin or Janice, but someone is married to Ralph.

> **THINKING AHEAD**
>
> How would we deal with the very basic claim that "Rodger is identical to Rodger"? Or how would we address a claim like "The morning star is actually the same as the evening star"? Do we have adequate tools to do so with LOLA as it presently stands?

## IDENTITY

*James Howlett and Wolverine are one and the same*

*h=o*

The final operator for LOLA is called the *identity* relation. Its symbol is the =, which reads "is identical to." The identity operator differs from all the other operators in that *it expresses a relation, not between statements, but between individuals*. Identity is, in fact, a 2-place predicate relation. It not only allows us to express the obvious relation that some individual is identical to itself, but also that two individuals are distinct, and that the same thing can be called by two names.

Now, you may be asking, why don't we just set up a translation scheme that includes Ixy: x is identical to y? We could, of course, do this. There are good reasons, however, for according identity its own symbol. First, unlike other predicates, identity expresses the *logical* truth that a thing is identical to itself. Second, a number of frequently occurring statement types, which we discuss below, require the identity relation to be adequately translated. The symbol shows immediately that these statements incorporate this logical relation.

Remember, the identity relation applies to individuals. So the symbol = is placed between individual constants or variables that are bound to quantifiers—NOT between statements:

$$a=b \qquad (x)(y)(Gx \rightarrow x=y)$$

We'll use the sentences in Example Set 6b to illustrate how we symbolize identity claims and negated identity claims.

## Example Set 6b
**TRANSLATION SCHEME**
RD: Everything

| b: Baroness Dudevant | g: George Sand | e: Venus | m: the morning star |
| --- | --- | --- | --- |
| n: the northern star | o: World War II | d: the war on drugs | |

1. Baroness Dudevant is identical to George Sand.
2. Venus is identical to the morning star.
3. The morning star is not the same as the northern star.
4. World War II is not identical to the war on drugs.
5. The morning star is the same thing as Venus.

These are all relatively straightforward symbolizations:

    **1. b=g**
    **2. e=m**
    **3. ~m=n**
    **4. ~o=d**
    **5. m=e**

In statements of non-identity such as 3 and 4, it is essential that you remember what they are stating. The tilde is operating over the *statement* of identity, it's *not* operating over either of the individuals. As we stated earlier, it would be meaningless to assert "the morning star is false" or "World War II is false." The tilde goes immediately before the identity statement, with no intervening parenthesis, just as for any 2-place predicate:

    **~a=b    ~Iab**

Quantified statements can also include identity relations. We hinted at the end of chapter five that identity is a 2-place predicate relation that is reflexive, symmetric, and transitive. We can demonstrate this now using the universal quantifier and the identity symbol.

Reflexive:    **(x)x=x**
Symmetric:   **(x)(y)(x=y → y=x)**
Transitive:    **(x)(y)(z)[(x=y & y=z) → x=z]**

Example Set 6c illustrates more kinds of statements that use quantifiers with identity.

### Example Set 6c
**TRANSLATION SCHEME**
RD: People

| p: the current president of the United States | c: the Chairman of the Board | Gx: x delivers the Gettysburg address |
|---|---|---|
| Sx: x sings *My Way* | Axy: x admires y | |

1. Anyone who's identical to the Chairman of the Board sang *My Way*.
2. Someone who's identical to the current president of the United States didn't deliver the Gettysburg address, but does sing *My Way*.
3. Everyone who is identical to herself admires herself.
4. Anyone who sings *My Way* is identical to either the current president of the United States or the Chairman of the Board.
5. Someone who admires the Chairman of the Board is not identical to the current president of the United States.

These are fairly straightforward statements to symbolize if you remember the basic patterns of universal and existential quantifications, and that *identity statements are just 2-place predicate statements*. For example, if we had included in our translation scheme Ixy: x is identical to y, you would symbolize statement 1 as: (x)(Ixc → Sx). To symbolize using the identity symbol, use the same pattern for universal quantifications, and just substitute x=c for Ixc.

$$1.\ (x)(x=c \rightarrow Sx)$$

Similarly for existential quantifications: The pattern for statement 2 is (∃x) [ __ & (__ & __)]. Using the identity symbol, we can fill in the pattern:

$$2.\ (\exists x)[x=p\ \&\ (\sim Gx\ \&\ Sx)]$$

*Please Note: do not put parentheses around simple identity statements. They are single statement units.*

With these points in mind, you should be able to symbolize the rest of the example statements. Take a moment to symbolize them, then compare your symbolizations with those below:

3. (x)(x=x → Axx)
4. (x)[Sx → (x=p v x=c)]
5. (∃x)(Axc & ~x=p)

## SOME, ALL, NONE, AND MORE

Up until this point, we have been somewhat limited in specifying quantities. To recap, if Px: x is a pine, and using the quantifiers we have, we can construct the following statements:

(x)Px: Everything is a pine.
(∃x)Px: Something is a pine.
~(x)Px: Not everything is a pine.
~(∃x)Px: Nothing is a pine.
(x)~Px: Everything is a non-pine.
(∃x)~Px: Something is not a pine.

The addition of identity, however, lets us symbolize statements that are more precise regarding quantity.

To develop these more complex statements, we need to pull together what we've learned about both existential and universal statements, identity statements, and our fluency in LOLA.

For example, consider the following,

There is at least one option. (RD: Everything; Ox: x is an option)

This is, of course, a very simple existential statement. The existential quantifier *means* that at least one thing *exists*, and that thing is an option. Our LOLA translation, then: (∃x)Ox. (There's at least one thing that is an option.)

Now consider this statement:

There are at least two options.

> There are at least two moons.
> (∃x)(∃y)[(Mx & My) & ~x=y]

Let's break down this English statement and sketch in the LOLA:

1) Two things exist [ **(∃x), (∃y)** ]
2) Those things are options [ **Ox & Oy** ]
3) The options are *different* options [ **~x=y** ]

From this breakdown, you can also construct a paraphrase of the original statement, and highlight the operators involved:

Two things **exist** such that **both** are options **and** they're **not** identical.

So, the LOLA:

(∃x)(∃y)[(Ox & Oy) & ~x=y]

Now, using this same pattern, you should quickly be able to translate the following into LOLA:

There are at least three options.

Note that as LOLA sentences becomes more complicated, you might want to sacrifice "parenthetical purity" for readability. Often it is easier to see what a WFF is saying if some non-essential parentheses are removed. Remember, though: *main connectives must still be unambiguously shown*. But often, as we discussed above, strings of ampersands or of wedges can be grouped together with little problem.

Below are both "parenthesis rich" and "parenthesis poor" versions of "There are at least three options":

(∃x)(∃y)(∃z) {[(Ox & Oy) & Oz] & [(~x=y & ~y=z) & ~x=z]}
(∃x)(∃y)(∃z)[(Ox & Oy & Oz) & (~x=y & ~y=z & ~x=z)]

Both translations unambiguously show that the main operator is the existential quantifier, and that the statement the existential quantifiers operate on is a conjunction.

---

**PRACTICE**

For this, as well as for the other patterns that follow, think up some of your own examples, and translate them into LOLA. Here, try translating:
There are at least four options.
There are at least five options.
There are at least six options.

You can see that (in theory and with unlimited time and paper) using this same pattern you could symbolize a statement such as, "There are at least 45 million people without health insurance." Or a computer could be programmed to give a LOLA translation of it. Not having unlimited time, we would never do this. But the point should be clear that using the identity relation allows us to say, in LOLA, much more than "Some uninsured people exist" or "Many uninsured people exist."

For the next three patterns, we'll use the following translation scheme. Again, for these patterns, we'll see that the identity relation allows us to be more precise as to quantity.

---

**Example Set 6d**
TRANSLATION SCHEME
RD: Everything

| Sx: x is a US state | Ix: x is a US island | Sxy: x is a senator from y |

| h: Hawaii | Hx: x is the 50th US state |

---

1. There is at most one US island state.

> There is at most one nugget.
> (x)(y)[(Nx & Ny) → x=y]

A quick look at this pattern might lead you to think that it is an existential quantification. But consider this situation: You're given a sack of five stones and you're told there is *at most* one gold nugget in the sack. Are you being told that a gold nugget is *actually* in the sack? No. If you found yourself holding a sack with five rocks, you couldn't say you'd been lied to. No one claimed that in fact *there is* a gold nugget.

So the first point is this: "at most" does not claim existence. And, since this pattern of statement does not claim existence, we don't use the existential quantifier. So how do we symbolize "at most" statements? Think about your bag of stones again. Say you cheat and look in the bag, and you find that there is a gold nugget. Now, reach in and pick up a stone. Great, you got the gold nugget. Now put the nugget back in, shake the bag, and pick up another stone. Hey, you got the gold again! Now: knowing that there was *at most* one gold nugget, but you picked a gold nugget twice, what *else* do you know? You know that the first and second nuggets must be the *same* nugget.

This is essentially how we logically analyze "at most" statements. For sentence 1, above, we'd analyze it as saying: For **any two** things we pick, **if** they are both U.S. island states, **then** they are identical. And our LOLA translation is:

**1. (x)(y){[(Ix & Sx) & (Iy & Sy)] → x=y}**

or the "parenthesis poor" version:

**1. (x)(y)[(Ix & Sx & Iy & Sy) → x=y]**

2. There are at most two senators from Hawaii.

Similarly for this statement: For **any three** things you pick, **if** they are all senators from Hawaii, **then two** of them must be identical. These are a bit trickier to symbolize, but just think about it: if two of them must be identical, then the 1st and 2nd **or** the 2nd and 3rd, **or** the 1st and 3rd *must be identical*. So, in LOLA:

**2. (x)(y)(z)[(Sxh & Syh & Szh) → (x=y v y=z v x=z)]**

3. There is exactly one US island state.

> There is exactly one nugget.
> (∃x)[Nx & (y)(Ny → x=y)]

For "exactly" statements, they do say that something exists, but they also limit how many there can be. If we told you there was exactly one gold nugget in the sack, you'd know there is one, but there can't be more than one. In other words, "exactly" statements are conjunctions of "at least" and "at most" statements, and so use both existential and universal quantifiers. For our example:

**3. (∃x)(Ix & Sx) & (x)(y)[(Ix & Sx & Iy & Sy) → x=y]**

We can also symbolize this in a more condensed form:

**3. (∃x){(Ix & Sx) & (y)[(Iy & Sy) → x=y]}**

4. There are exactly two senators from Hawaii.

Again, for this statement, it states that there are two senators from Hawaii, and they are limited to two senators.

**4. (∃x)(∃y){(Sxh & Syh & ~x=y) & (z)[Szh → (z=x v z=y)]}**

5. The 50th US state is an island.

Sentence 5 contains a very interesting kind of term called a "definite description." We've seen such descriptions before: in Example Sets 6b and 6c, above, we treated such descriptions, essentially, the same as names, and we symbolized them with individual constants:

> The author of *Hamlet* is British.
> (∃x)[(Hx & Bx) & (y)(Hy → y=x)]

m: the morning star
p: the current president of the US

There seems to be some justification for this: definite descriptions give a precise description that purports to pick out exactly one entity, as in the claim "The current president of the USA is a Democrat."

But this is a controversial way to represent definite descriptions. Another way of looking at the matter is to analyze the entire statement that contains the definite description. On this view, the term "is the 50th US state" is a predicate! And Hawaii does seem to have the *property* of being the 50th US state, although, unlike most predicates, it is the only thing that does have this property.

Put more formally, the philosopher Bertrand Russell argued that sentences like **5** are actually asserting three things:

1) A particular thing exists (the 50th US state).
2) This thing possesses a particular trait (is an island).
3) There is only one of them, that is, it's unique.

The first component would be: (∃x)Hx
The second component would be: Ix
The third component would be: (y)(Hy → y=x)

Then put them all together, making sure that the final x is bound to the initial existential quantifier.

**5. (∃x)[(Hx & Ix) & (y)(Hy → y=x)]**

So, are definite descriptions like individual constants or are they more like predicates? In the spirit of trying to show as much logical detail as we can, we will treat sentences like **5** as Russell did.

For the last three patterns, we'll use the following translation scheme:

## Example Set 6e
**TRANSLATION SCHEME**
RD: People

| b: Humphrey Bogart | r: Rick | c: *Casablanca* |
| --- | --- | --- |
| Pxyz: x plays y in z | Sx: x is a male superstar | Ox: x wins an Oscar |
| Fxy: x is more famous than y | | |

1. Only Bogart could play Rick in *Casablanca*.

> The only flying elephant is Dumbo.
> (Ed & Fd) & (x)[(Ex & Fx) → x=d]

"Only" statements can be phrased in more than one way. Variants of sentence **1** are

The only one who could play Rick in *Casablanca* is Bogart.
No one could play Rick in *Casablanca* except Bogart.

These statements are fairly straightforward. They contain at least

A predicate statement (an individual with a 1, 2, 3, or n-place predicate).
A universal statement that asserts anything (person, etc.) who has that predicate is identical to the individual in the predicate statement.

In other words, statement **1** can be paraphrased as:

Bogart could play Rick in *Casablanca* **and for any** person, **if** that person could play Rick in *Casablanca*, **then** that person is identical to Bogart.

Translated to LOLA, we get:

**1. Pbrc & (x)(Pxrc → x=b)**

2. The only male superstar who could play Rick in *Casablanca* is Bogart.

This "only" statement is similar to statement **1**. The difference here is that instead of one predicate statement, we see there are two predicates we can apply to Bogart. So our LOLA sentence is:

**1. (Sb & Pbrc) & (x)[(Sx & Pxrc) → x=b]**

3. All male superstars except Bogart won an Oscar.

"All ... except" statements are also conjunctions. Let's look closely at all the information contained in this one sentence, and we'll sketch in the LOLA as we go. We find that this one sentence contains three statements.

> All students except Smith will pass.
> (Ss & ~Ps) & (x)[(Sx & ~x=s) → Px]

Bogart is a male superstar. Sb
Bogart didn't win an Oscar. ~Ob
Any person, if that person is a male superstar and is not identical with Bogart, then he did win an Oscar. (x)[(Sx & ~x=b) → Ox]

Putting it all together, we get:

**1. (Sb & ~Ob) & (x)[(Sx & ~x=b) → Ox]**

4. Bogart is the most famous male superstar.

Sentences such as 4 contain *superlatives*. Before we get to the logic of statements of this pattern, let's do a quick grammar review. Adjectives and

> Ali is the greatest fighter.
> Fa & (x)[(Fx & ~x=a) → Gax]

adverbs come in degrees called positive, comparative, and superlative. The chart below gives some examples.

| **Positive** | Good | Rich | Quickly | Famous |
|---|---|---|---|---|
| **Comparative** | Better | Richer | More quickly | More famous |
| **Superlative** | Best | Richest | Most quickly | Most famous |

We've often seen *comparative* forms in many of the exercises we've worked; they were two-place predicates, such as:

Gxy: x is greater than y          Mxy: x is more real than y       Oxy: x is older than y
Vxy: x is more violent than y     Nxy: x is nicer than y           Sxy: x is smaller than y

And in our current translation scheme: Fxy: x is more famous than y. As we'll see in the analysis below, we use the *comparative* form as part of the *superlative* expression.

Superlative statements, like the two preceding kinds of statements, talk about "one of a kind" individuals. But further, superlatives pick out what we could call "best in class" individuals. To say that "Voldemort is the most evil wizard" means that no matter what *other* wizard you compare him to, Voldemort will be more evil. (Be careful: we don't want to say "No matter *what wizard* you compare him to, Voldemort will be more evil," because Voldemort is a wizard, so that statement implies that he's more evil than himself. He's evil, but nobody can be *that* evil.)

And this is essentially how we analyze superlative statements. For statement **4**, the superlative statement claims:

Bogart is a male superstar.

For ***any*** person, ***if*** they're a male superstar ***and not identical*** with Bogart, ***then*** Bogart is more famous than he is.

And in LOLA,

**1. Sb & (x)[(Sx & ~x=b) → Fbx]**

This brings us to the end of our discussion of identity statements. Below is a table that summarizes the *basic* patterns of the last seven statement types.

| At least | At most |
|---|---|
| There are at least two options. (∃x)(∃y)(Ox & Oy & ~x=y) | There is at most one option. (x)(y)[(Ox & Oy) → x=y] |

| Exactly | Definite description |
|---|---|
| There is exactly one option. (∃x)[Ox & (y)(Oy → x=y)] | The singer of *My Way* is from New Jersey. (∃x)[(Sx & Jx) & (y)(Sy → y=x)] |

| The only | All ... except |
|---|---|
| Only Elizabeth II is Queen of England. Qe & (x)(Qx → x=e) | All farmers except Jones lost crops. (Fj & ~Cj) & (x)[(Fx & ~x=j) → Cx] |

| Superlative | |
|---|---|
| The most famous dog is Lassie. Dl & (x)[(Dx & ~x=l) → Flx] | |

# Exercise 6b

Symbolize the following statements into LOLA.

## TRANSLATION SCHEME
RD: People

| e: Einstein | j: Joseph | h: Harold | c: Carol | Kxy: x knows about y |
| Mx: x is a math student | Sx: x is a science student | Bxy: x is a better science student than y | Txy: x teaches y | Uxy: x tutors y at Mainland High School |

1. Harold tutors some of the math students at Mainland High School.
2. Given that Joseph tutors most of the science students at Mainland, he would know about Einstein.
3. Einstein never taught math students, but he studied math.
4. Someone taught Einstein.
5. Either Einstein knew about Harold or Harold knew about Einstein.
6. Being a teacher is an asymmetric relationship.
7. Carol is the best science student.
8. The only science student is Carol.
9. There are exactly two math students.
10. Every science student except Carol tutors Harold at Mainland High School.

# Exercise 6c

Using the translation scheme below, symbolize the following statements into LOLA.

## TRANSLATION SCHEME
RD: People

| b: Batman | r: Bruce Wayne | j: Dr. Jekyll | h: Mr. Hyde |
| Fx: x fights crime | Gx: x is guilty | Cx: x is a caped crusader | Fxy: x is more famous than y |

1. Batman is identical to Bruce Wayne.
2. Dr. Jekyll is identical to Mr. Hyde.
3. Batman is not identical to Dr. Jekyll.
4. Bruce Wayne is not identical to Mr. Hyde.
5. Anyone identical to Batman fights crime.
6. There is someone identical to Mr. Hyde who is guilty.
7. Batman is the only caped crusader.
8. Everyone except Batman is a caped crusader.
9. All guilty crime fighters are identical to Dr. Jekyll.
10. There is exactly one caped crusader.

11. People are caped crusaders if and only if they fight crime.
12. A person is identical to Batman if and only if that person fights crime.
13. At least two people are guilty.
14. At most two people are guilty.
15. Exactly two people are guilty.
16. Batman is the most famous caped crusader.
17. Being a caped crusader is a necessary and sufficient condition for fighting crime.
18. "Being more famous than" is a transitive relationship.
19. In order to be a non-guilty, crime-fighting, caped crusader, one must be identical to Batman.
20. Although Mr. Hyde is the most famous guilty caped crusader, not all guilty caped crusaders are identical to Mr. Hyde.

## Exercise 6d

Create a translation scheme and then symbolize the following statements into LOLA.

1. Tarantino makes exciting movies, but so does Hitchcock.
2. Tarantino won't make an exciting movie unless Hitchcock does.
3. Wood will win an Oscar if he makes an exciting movie.
4. Wood will win an Oscar only if some director doesn't win one.
5. Wood is better than Hitchcock unless Hitchcock is better than Tarantino.
6. Hitchcock admires no directors, but many directors do admire Hitchcock.
7. Ed Wood won't win an Oscar unless neither Hitchcock nor Tarantino do.
8. "Admiring" is a non-transitive relation.
9. Some Oscar winning directors admire some directors who make exciting movies.
10. Not all directors who admire Hitchcock admire Tarantino.
11. Although no directors admire Ed Wood, he does make exciting movies.
12. Tarantino is a better director than Wood, but Hitchcock is the best director.
13. Either Wood or Hitchcock won an Oscar, but not both.
14. Neither Hitchcock nor Wood won an Oscar, yet some directors who make exciting movies are Oscar winners.
15. The only Oscar winning director is Tarantino.
16. Hitchcock is the best Oscar winning director.
17. There are at least two Oscar winning directors.
18. All Oscar winning directors except Wood make exciting movies.
19. All directors are better than Hitchcock if and only if Hitchcock is not better than Tarantino.
20. All directors who are identical with Wood don't make exciting movies.

# CHAPTER SIX: BICONDITIONALS AND IDENTITY

## THINKING AHEAD

Consider what it means to say that a statement is true or false. Now what would it mean if you combined two statements with a conjunction, one of them false and one of them true? Would the resulting statement be true or false? How would you assess a disjunction made up of two false statements? Or a disjunction with one true statement and one false statement?

# Unit One Review

| Vocabulary of LOLA | |
|---|---|
| **Individual Constants (names)**<br>a – t<br>**Individual Variables**<br>u – z<br>**Predicate Letters**<br>A – Z<br>**Parentheses**<br>( ), [ ], { }<br>**Statement Letters**<br>A – Z | **1-place Operators**<br>Quantifiers: (x), (∃y)<br>Tilde: ~<br>**2-place Operators**<br>Ampersand: &<br>Wedge: v<br>Arrow: →<br>Double arrow: ↔<br>Identity: = |

| For Grammatically Correct Sentences in LOLA |
|---|
| **RULE 1:** Predicates, names, and variables never stand alone.<br>  Predicates stand with names or bound variables  (Ex. Pa; (x)Pxb)<br>  Names stand with a predicate (incl. =)<br>  Variables always bound and stand with predicates (incl. =) |
| **RULE 2:** 1-place operators go immediately before a grammatically correct sentence in LOLA. |
| **RULE 3:** The &, v, →, and ↔ are placed between 2 grammatically correct sentences in LOLA. |
| **RULE 4:** The = is placed between two names or bound variables. |
| **RULE 5:** Parentheses enclose quantifiers or group two grammatically correct sentences in LOLA. |
| **RULE 6:** Every grammatically correct sentence in LOLA *with* operators has a *main* operator. |

## Common English Variations

### Universal Quantifiers
All, every, each, each and every

### Negations
It's false (not true) that, not, un-, dis-, mis-

### Conjunctions
And, but, yet, although, even though, however, nevertheless, which

### Disjunctions
Or, either or, and/or, unless

### Existential Quantifiers
Some, many, most, a few, there is a, there exists, at least one is, there are

### Conditionals
If ... then, only if, not ... unless, necessary condition for, sufficient condition for

### Biconditionals
If and only if, just in case, necessary and sufficient condition for, is defined as

### Identity
Is identical to, is the same person (creature, entity) as

## LOLA Alternatives

| | |
|---|---|
| Not ... unless<br>$\_ \rightarrow \_$<br>$\sim\_ \text{ v } \_$ | Not every ..., Some are not ...<br>$\sim(x)$<br>$(\exists x)\sim$<br>$(\exists x)(\_x \ \& \sim\_x)$ |
| Neither ... nor<br>$\sim(\_ \text{ v } \_)$<br>$\sim\_ \ \& \ \sim\_$ | There are no, No ... is (are, has, etc.)<br>$\sim(\exists x)$<br>$(x)\sim$<br>$(x)(\_x \rightarrow \sim\_x)$ |

## Important Relations

| Relation | Reflexivity 1 | Symmetry 2 | Transitivity 3 |
|---|---|---|---|
| "Positive" | Reflexive<br>(x)Rxx | Symmetrical<br>(x)(y)(Rxy → Ryx) | Transitive<br>(x)(y)(z)[(Rxy & Ryz) → Rxz] |
| "Negative" | Irreflexive<br>(x)~Rxx | Asymmetrical<br>(x)(y)(Rxy → ~Ryx) | Intransitive<br>(x)(y)(z)[(Rxy & Ryz) → ~Rxz] |
| "Neutral" | Nonreflexive<br>(x)(Rxx v ~Rxx) | Nonsymmetrical<br>(x)(y)[Rxy → (Ryx v ~Ryx)] | Nontransitive<br>(x)(y)(z)[(Rxy & Ryz) → (Rxz v ~Rxz)] |

# Unit One: Answers to Selected Problems

**CHAPTER ONE**

**Exercise 1a**

3. Nobody's perfect.
   **A perfect person does not exist.**
   **Everyone is flawed.**

4. Don't we say "thank you" when someone gives us a present?
   **You should thank (Aunt So-and-So or whomever) for your present.**
   **Say "thank you" for the gift.**

6. Her auburn tresses, glittering in the sun, lured an infestation of migrating drones.
   **Bees attacked the girl's sun-lit red hair.**
   **That carrot-top woman is attracting bees.**

**CHAPTER TWO**

**Exercise 2a**

TRANSLATION SCHEME

| b: Bob | s: Susan | e: the *Queen Elizabeth* | t: the *Titanic* | r: the Roaring Twenties |
|---|---|---|---|---|
| Mx: x moves quickly | Wx: x wants a great deal | Tx: x is tall | Bx: x is big | Dx: x is dangerous |

1. Bob moves quickly. Mb
4. Susan wants a great deal. Ws
7. The *Queen Elizabeth* is big. Be
8. The *Titanic* was tall. Tt
11. The Roaring Twenties were dangerous. Dr

**Exercise 2b**

1. Bf  Frank is beautiful.
2. Sm  Mary is quite chic.
3. Da  The ambassador is boring.
8. Ls  San Francisco loves a good time.

## Exercise 2c

**TRANSLATION SCHEME**

| d: Detroit | n: Newark | r: Roberto | t: Tom |
|---|---|---|---|
| c: Chicago | Sxy: x is smaller than y | Nxy: x is nicer than y | Jxy: x is jealous of y |

3. Detroit is smaller than Newark. **Sdn**
5. Newark is nicer than Chicago. **Nnc**
7. Tom is bigger than Roberto. **Srt**
13. Newark is jealous of Detroit. **Jnd**

## Exercise 2d

1. Eoo   **One is equal to one.**
3. Dnt   **Ten is evenly divisible by two.**
4. Ltl   **Two is less than twelve.**
14. Sfi   **Four is succeeded by five.**

## Exercise 2e

**TRANSLATION SCHEME**

| a: Adam Smith | l: Plato's nose | b: Barbra Streisand | s: Barbra Streisand's nose | g: Groucho Marx |
|---|---|---|---|---|
| k: Karl Marx | Sx: x has a snub nose | Bxy: x is bigger than y | Gx: x is a genius | Sxy: x is smarter than y |

1. Adam Smith has a snub nose. **Sa**
2. Plato's nose is bigger than Barbra Streisand's. **Bls**
4. Groucho Marx is a genius. **Gg**
5. Karl Marx is smarter than Barbra Streisand. **Skb**

## Exercise 2f

1. Vsb   **Swimming is more violent than baseball.**
2. Mh   **Hockey is monotonous.**
3. Ds   **Swimming is deadly.**
5. Bfl   **Football is more boring than lacrosse.**

## Exercise 2g

2. Ilfs **Lithuania imports more from France than from the United States.**
3. Fscl **The United States is farther from Canada than Lithuania is.**
8. Tcsf **Canada talks with the United States more than with France.**
10. Lflc **France likes Lithuania more than Canada does.**
17. Ifsc **France imports more from the United States than from Canada.**

## Exercise 2h

TRANSLATION SCHEME

| c: Cathy | d: David | m: Chicago Museum of Art | n: New York Museum of Modern Art | l: LA Getty Museum |
|---|---|---|---|---|
| s: Susan | o: the *Mona Lisa* | Bxyz: x is between y and z | Exyz: x brought y to z | Gxyz: x gives y to z |

2. The Chicago Museum of Art is between the New York Museum of Modern Art and the Los Angeles Getty Museum. **Bmnl**
3. Susan brought David to the New York Museum of Modern Art. **Esdn**
9. Cathy gave the *Mona Lisa* to David. **Gcod**
14. The Los Angeles Getty Museum gave the *Mona Lisa* to the New York Museum of Modern Art. **Glon**
15. The *Mona Lisa* is between The New York Museum of Modern Art and the Chicago Museum of Art. **Bonm**

## Exercise 2i

1. One is between two and three. **Both**
2. Four succeeds one by three. **Sfoh**
3. Three is odd. **Oh**
4. Four is even. **Ef**
5. Four is less than ten by six. **Lfns**

# CHAPTER THREE

## Exercise 3a

*Set 1*

**TRANSLATION SCHEME**
RD: People

| b: Bob | s: Susan | Bx: x is bored | Fxy: x likes to fly more than y |
|---|---|---|---|

1. Somebody is bored. **(∃x)Bx**
3. Bob likes to fly more than some people. **(∃z)Fbz**
5. At least one person likes to fly more than Susan. **(∃w)Fws**
7. Bob likes to fly more than Susan. **Fbs**

## Exercise 3b

*Set 2*

1. (∃x)Txmk    **Somebody telephoned Mary about Karen.**
3. (∃z)Lbz     **Bob likes somebody.**
4. (∃y)Tmby    **Mary will telephone Bob about somebody.**

## Exercise 3c

*Set 3*

1. All philosophers are wise. **(x)Wx**
2. One philosopher is smarter than all philosophers. **(∃x)(y)Sxy**
3. All philosophers teach logic to some philosophers. **(x)(∃y)Txy**
4. Aristotle teaches logic to all philosophers. **(x)Tax**
5. All philosophers pass their wisdom along to all philosophers via some philosopher. **(x)(y)(∃z)Pxyz**

*Set 4*

2. (∃y)Ny       **Somebody's nagging.**
3. (∃x)Lmxs     **My mom wants someone to love my sister.**
8. (∃x)(y)Fxy   **Somebody's fighting with everyone.**
9. (y)(∃z)Fyz   **Everyone fights with somebody.**
10. (∃z)Lszb    **My sister wants somebody to love my brother.**

## Exercise 3d

TRANSLATION SCHEME

RD: Cars

| d: Dan's Suzuki | l: Lane's Honda | p: the President's limo | Fxy: x is faster than y | Pxyz: x parks between y and z |
|---|---|---|---|---|

1. Any car is faster than Dan's Suzuki. **(x)Fxd**
6. The President's limo is faster than every car. **(x)Fpx**
7. At least one car is parked between the President's limo and Dan's Suzuki. **(∃x)Pxpd**
9. At least one car is faster than Lane's Honda. **(∃x)Fxl**

## CHAPTER FOUR

### Exercise 4a

3. The sun isn't friendlier than the earth. **~Fse**
6. It's not unhot on the sun. **~~Hs**
7. The sun causes the moon to affect the earth's orbit. **Csme**
8. The moon doesn't cause the sun to affect the earth's orbit. **~Cmse**
9. The earth doesn't cause the moon to affect the sun's orbit. **~Cems**
15. ~Cjse    **Jupiter doesn't cause the sun to affect the earth's orbit.**
16. Fjm    **Jupiter is friendlier than the moon.**
18. Cspj    **The sun causes Pluto to affect the orbit of Jupiter.**
19. ~Fpj    **Pluto isn't friendlier than Jupiter.**
20. ~Hs    **It's false that it's hot on the sun.**

### Exercise 4b

TRANSLATION SCHEME

| d: Detroit | c: Chicago | l: Los Angeles | n: New York |
|---|---|---|---|
| Dx: x is dull | Cxy: x is close to y | Bxyz: x is between y and z | Fx: x has lots of factories |

2. Chicago is not close to Los Angeles. **~Ccl**
3. Los Angeles isn't between Chicago and New York. **~Blcn**
6. New York doesn't have lots of factories. **~Fn**
8. Chicago isn't dull. **~Dc**
9. Los Angeles is close to Detroit. **Cld**

## Exercise 4c

1. ~(x)Pxy **negation**
3. (∃y)(x)Mxy **existential quantification**
5. ~~(x)~Pxm **negation**
6. (x)~(∃y)Mxy **universal quantification**

## Exercise 4d

1. Nothing is popular. **~(∃x)Px OR (x)~Px**
3. Some things are not popular. **(∃x)~Px OR ~(x)Px**
6. Nothing is more popular than Santa Claus. **~(∃x)Mxs OR (x)~Mxs**
14. It's false that everything is unpopular. **~(x)~Px OR (∃x)Px**
15. Everything is unpopular. **(x)~Px OR ~(∃x)Px**

## Exercise 4e

1. Gj **John is a gossip.**
2. ~Gs **Susan isn't a gossip.**
3. (∃y)~Gy **Somebody isn't a gossip.**
4. ~(x)Gx **Not all people gossip.**
14. ~(∃z)Gz **No gossips exist.**

## Exercise 4f

**TRANSLATION SCHEME**
RD: Birds

| p: the penguin at the zoo | Wx: x has wings | Fxy: x is faster than y |
|---|---|---|

3. Not all birds have wings. **~(x)Wx OR (∃x)~Wx**
4. The penguin at the zoo does not have wings. **~Wp**
6. The penguin at the zoo is not faster than some birds. **(∃x)~Fpx**
7. All birds are faster than some bird. **(x)(∃y)Fxy**
9. Not every bird is faster than all birds. **(∃x)(y)~Fxy**

## Exercise 4g

1. Hannibal is dangerous and smart. **Dh & Sh**
2. It's not true that everyone is dangerous as well as smart.
   **(∃x)[Px & ~(Dx & Sx)]**
3. Some electrical devices are dangerous. **(∃x)(Ex & Dx)**
4. Some people are not dangerous. **(∃x)(Px & ~Dx)**

# CHAPTER FIVE

## Exercise 5a

1. Everyone who likes coffee is jittery. **(x)(Cx → Jx)**
3. Some people will be happy only if everyone is unhappy.
   **(∃x)Hx → (x)~Hx**
5. No one who likes tea likes coffee. **(x)(Tx → ~Cx)**
8. Everyone likes coffee if Allen does. **Ca → (x)Cx**
11. (∃z)Tz → (∃x)~Cx   **If someone likes tea, then there are people who don't like coffee.**
14. (z)(Cz → ~Tz)   **Everyone who likes coffee doesn't like tea.**
15. ~(v)(Cv → ~Hv)   **Not everyone who likes coffee is unhappy.**
20. (x)Tx → (∃z)(~Cz & ~Jz)   **If everyone likes tea, then there're people who don't like coffee, but they're not jittery.**

## Exercise 5b

TRANSLATION SCHEME
RD: Everything

| r: Rover | Ax: x is alive | Ox: x needs oxygen |
|---|---|---|
| Rx: x reproduces | Mx: x mates | |

1. Everything that is alive needs oxygen. **(x)(Ax → Ox)**
3. Some things reproduce, but don't mate. **(∃x)(Rx & ~Mx)**
8. If some things mate, then at least one thing is alive. **(∃x)Mx → (∃y)Ay**
11. Not everything that needs oxygen is alive. **~(x)(Ox → Ax)   OR   (∃x)(Ox & ~Ax)**
20. If Rover is not alive, then something is not alive. **~Ar → (∃x)~Ax**

## Exercise 5c

2. If someone boring wrote *War and Peace*, then it's a boring book.
   **(∃x)[(Px & Bx) & Wxa] → (Ta & Ba)**
6. If someone read *War and Peace*, then somebody read an interesting book.
   **(∃x)(Px & Rxa) → (∃x)(∃y){[Px & (Ty & Iy)] & Rxy}**
8. Someone is a boring author, but not Camus.
   **(∃x)[Px & (Ax & Bx)] & [Pc & (Ac & ~Bc)]**
9. If anyone loans a book to Tolstoy, then that person will get his book back.
   **(x)(y){[(Px & Ty) & Lxyt] → Gx}**
16. If anyone loans a boring book to someone, then they'll not get their book back. **(x)(y)(∃z)({[(Px & Ty) & (By & Pz)] & Lxyz} → ~Gx)**

24. (x)[(Ax & Ix) → (Ix & Px)]  **Any interesting author is an interesting person.**
25. Wdp v (∃x)[(Px & Ax) & Wxp]  **Either Dostoevsky wrote *Crime and Punishment* or someone who's an author wrote it.**
26. (x)(∃y){[Px & (Ty & By)] → Rxy}  **Every person reads some boring books.**
29. (x)[(Ix & Px) → {(Bx & Ax) v (~Bx v Ax)}]  **Every interesting person is either a boring author or not boring unless an author.**
35. (∃x){Px & (∃y)[Ty & (Lxyd v Lxyc)]}  **Somebody loaned some book to either Dostoevsky or Camus.**

### Exercise 5d

1. "Being identical to" is a reflexive relation. **(x)Ixx**
5. "Being the mother of" is an asymmetrical relation. **(x)(y)(Mxy → ~Myx)**
9. "Loving" is a nontransitive relation. **(x)(y)(z)[(Lxy & Lyz) → (Lxz v ~Lxz)]**

### Exercise 5e

2. Albert and Burt are partners. **Pab & Pba**
6. Carmen will go to college if Albert or someone goes. **[Ca v (∃x)Cx] → Cc**
8. No one will go to college unless both Albert and Burt go. **~(∃x)Cx v (Ca & Cb)**
9. If anyone goes to college, Carmen will go. **(∃x)Cx → Cc**
10. Burt's the ancestor of Carmen or the ancestor of Albert, but not both. **(Abc v Aba) & ~(Abc & Aba)**

## CHAPTER SIX

### Exercise 6a

2. A necessary and sufficient condition for being a bachelor is being an unmarried adult male. **(x){Bx ↔ [~Wx & (Ax & Mx)]}**
5. It's untrue that "being happier than" is a symmetrical relation. **~(x)(y)(Hxy → Hyx)**
6. Ralph is a happy bachelor if and only if he is a happy, adult, unmarried male. **(Hr & Br) ↔ [(Hr & Ar) & (~Wr & Mr)]**
9. Janice is happier than everyone if and only if she's not married. **(x)(Hjx ↔ ~Wj)**
15. Every single adult male is a bachelor, but even though Ralph is an unmarried male, he's not an adult. **(x){[~Wx & (Ax & Mx)] → Bx} & [(~Wr & Mr) & ~Ar]**

## Exercise 6b

3. Einstein never taught math students, but he studied math.
   **(x)(Mx → ~Tex) & Me**
7. Carol is the best science student. **Sc & (x)[(Sx & ~x=c) → Bcx]**
8. The only science student is Carol. **Sc & (x) (Sx → x=c)**
9. There are exactly two math students.
   **(∃x)(∃y){(Mx & My & ~x=y) & (z)[Mz → (z=x v z=y)]}**
10. Every science student except Carol tutors Harold at Mainland High School. **(Sc & ~Uch) & (x)[(Sc & ~x=c) → Uxh]**

## Exercise 6c

1. Batman is identical to Bruce Wayne. **b=r**
3. Batman is not identical to Dr. Jekyll. **~b=j**
5. Anyone identical to Batman fights crime. **(x)(x=b → Fx)**
6. There is someone identical to Mr. Hyde who is guilty. **(∃x)(x=h & Gx)**
7. Batman is the only caped crusader. **Cb & (x)(Cx → x=b)**
8. Everyone except Batman is a caped crusader. **~Cb & (x)(~x=b → Cx)**
10. There is exactly one caped crusader. **(∃x)Cx & (x)(y)[(Cx & Cy) → x=y]**
14. At most two people are guilty.
    **(x)(y)(z)[(Gx & Gy & Gz) → (x=y v y=z v x=z)]**
16. Batman is the most famous caped crusader.
    **Cb & (x)[(Cx & ~x=b) → Fbx]**

## Exercise 6d

**TRANSLATION SCHEME**
RD: Directors

| t: Tarantino | h: Hitchcock | e: Ed Wood | Ex: x makes exciting movies |
|---|---|---|---|
| Ox: x wins an Oscar | Bxy: x is better than y | Axy: x admires y | |

6. Hitchcock admires no directors, but many directors do admire Hitchcock.
   **~(∃x)Ahx & (∃x)Axh**
9. Some Oscar winning directors admire some directors who make exciting movies. **(∃x)(∃y)[(Ox & Ey) & Axy]**
12. Tarantino is a better director than Wood, but Hitchcock is the best director. **Bte & (x)(~x=h → Bhx)**
15. The only Oscar winning director is Tarantino. **Ot & (x)(Ox → x=t)**
18. All Oscar winning directors except Wood make exciting movies.
    **(Oe & ~Ee) & (x)[(Ox & ~x=e) → Ex]**
20. All directors who are identical with Wood don't make exciting movies.
    **(x)(x=e → ~Ex)**

# UNIT TWO: IS IT TRUE?

# Properties and Relations of Statements

# Chapter Seven: Connectives and Truth Tables

*What's Up?*
LOLA-Lite
Truth-functional Connectives
Truth Tables

Throughout the first six chapters we have been analyzing the sometimes very complex patterns of statements. As we saw with the introduction of certain operators—the tilde (~), the conjunction (&), the disjunction (v), the conditional (→), and the biconditional (↔), statements are frequently made up of more than a single claim. *These operators are often called connectives* because they *connect* one statement to another (or, in the case of the tilde, a statement to a denial). So, often a statement is a combination of several claims that form a complex statement. In order to clarify this we make a distinction between *atomic* and *compound* statements. *Whenever you have a statement that has no connectives, it's called an atomic statement. A compound statement has at least one connective.*

This raises an interesting issue with regard to the nature of statements when we want to consider *under what conditions we call a statement true or false*. It's clear how we'd classify a statement composed of nothing but an individual and a predicate as true or false (though sometimes we're not certain about the facts and can't confidently make this classification). For example, the statement, "George Washington was the first president of the United States" is true and the statement "George Washington was born in 1983" is false. However, what should we do with a compound statement where there is a combination of statements and one of the parts is false and the other part is true? Is the statement *as a whole* true or false?

The answer is that it *depends upon which connective is involved*. In the next three chapters we will go over the different ways that connectives can affect the overall truth status of a proposition and how these operators create special classes of statements. To do this we introduce a new way of translating statements and an important technique in LOLA called truth tables.

## CONNECTIVES AND A NEW METHOD OF SYMBOLIZING: LOLA-LITE

The connectives are the wedge, the tilde, the arrow, the double arrow, and the ampersand. In addition to what we learned in previous chapters about how each of these connectives translates different English expressions, they each have the additional property of being "truth functional." This means that *the overall truth status of a compound statement is a function of the truth status of the component atomic statements and the rules governing its connectives*. This will be easier to understand when we've demonstrated a couple of them. We'll begin with the tilde by considering the following sentence.

It's not true that George Washington was born in Thailand.

Normally, you would construct a translation scheme that would allow you to represent the individual "George Washington," the individual "Thailand," and the relationship of "x being born in y"

g: George Washington
t: Thailand
Bxy: x was born in y

then you'd symbolize it,

**Bgt**

then you'd place the tilde to the left of it,

**~Bgt**

If our primary concern, however, is the truth-value of a compound statement, we can shift our focus from the *internal* details of the constituent atomic statement (in this case, Bgt) to the connective. Often, all we need to know is how many different simple statements make up the compound, and what connectives hold them together. We can use a simplified language, LOLA-Lite, to clearly show the "bare bones" of the compound.

In LOLA–Lite, a single capital letter stands for a single atomic statement. Thus, we can symbolize "George Washington was born in Thailand" simply as

**G**

If we stopped here, you'd have an *atomic* statement. However, we're not done yet. We need to put the tilde to its left, giving us,

**~G**

As we said above, because there is a connective involved, this is a *compound* statement. You have an atomic statement "G" and then by adding the operator, the tilde, you create a compound statement. By definition, all compound statements include at least one atomic statement.

> *Alaska is the largest state and New York is the largest city in the USA.*
> S & C
> What would this be in full LOLA?

This ability to shift from one level of analysis to another, depending upon the task involved is a fundamental skill in logic. Sometimes all you want to focus on is whether a particular statement is a conditional or a disjunction; you don't need to worry about whether its components are quantifiers or identity statements as well. Moreover, as we said earlier, this will also allow us to introduce a logical technique called a truth table which, while extremely useful, is pretty much limited to using this new method of translation, LOLA-Lite.

For the next three chapters you'll translate the statements without predicates, quantifiers, or identity statements, in order to practice this new type of translation. However, to make sure that you don't lose the skills that you've developed so far, we'll often ask you to translate a set of practice exercises in both this more attenuated version and the more complete method you've already mastered. In addition, we'll include translation exercises which are devoted just to translation practice with the full-LOLA approach we covered in Chapters Two through Six.

## Exercise 7a

Symbolize the following statements using the translation schemes provided at the end of each sentence.

1. Fido ran into the street. (F: Fido ran into the street.)
2. It's not true that Fido ran into the street. (F: Fido ran into the street.)
3. John's cat is overweight and spoiled. (O: John's cat is overweight; S: John's cat is spoiled.)
4. Bob's cat is neither overweight nor spoiled. (O: Bob's cat is overweight; S: Bob's cat is spoiled.)
5. John doesn't like cats or dogs. (D: John likes dogs; C: John likes cats.)
6. Assuming John doesn't like dogs, he won't like Fido. (D: John likes dogs; F: John likes Fido.)
7. Sam will buy John a dog only if he promises to take care of it. (B: Sam buys John a dog; P: John promises to take care of the dog.)
8. Lorie won't graduate this term unless she finishes her coursework. (G: Lorie graduates this term; F: Lorie finishes her coursework.)

9. One will have a fire if and only if there is fuel, oxygen, and an ignition of some sort. (F: There is a fire; S: There is fuel; O: There is oxygen; I: There is ignition of some sort.)
10. People don't go to the theater very often, but tonight some will. (T: People often go to the theater; G: Tonight some go to the theater.)
11. The theater will be crowded tonight but safe, if the security people know what they're doing. (C: The theater is crowded tonight; S: The theater is safe; K: The security people know what they're doing.)
12. That is a buffalo just in case it is a bison. (U: That is a buffalo; I: That is a bison.)
13. Most people complain about the weather, although none of them are willing to do anything about it. (C: People complain about the weather; W: People are willing to do something about the weather.)
14. Either a person takes the test or they don't; they can't have it both ways. (T: A person takes the test.)
15. John's having a severe medical condition is a sufficient condition for his being discharged from the army. (M: John has a severe medical condition; D: John is discharged from the army.)
16. The undergraduate college, but not the graduate school, is committed to ending student plagiarism. (U: The undergraduate college is committed to ending student plagiarism; G: The graduate school is committed to ending student plagiarism.)
17. If it rains this Sunday, then the picnic will not be held, but people can check the website or call John to make sure. (R: It rains this Sunday; P: The picnic will be held; C: People can check the website to make sure; T: People can call John to make sure.)
18. Passing at least three of the four tests is a necessary condition for passing this course. (P: A person passes this course; T: A person passes three of the four tests.)
19. The first medical tests that have been done are enough to determine Lorie's condition, but no one will know how to treat her unless the doctors do some more. (M: We have done the first medical tests; C: We can tell Lorie's condition; K: The doctors know how to treat Lorie; N: The doctors need to do more medical tests.)
20. People are going to fund the project only if they like the script and they believe that you can meet all the deadlines. (F: People fund the project; L: People like the script; B: People believe you can meet all the deadlines.)

> **KEEP IN PRACTICE**
>
> Create a complete translation scheme (names, quantifiers, predicates, etc.) for the above sentences, and then translate the sentences into LOLA.

## Exercise 7b

Using the following translation scheme, translate the symbolic statements from LOLA-Lite into English.

### TRANSLATION SCHEME

| A: All of us are busy. | C: We can meet you at 10:00. | E: Evan wants to join us. | L: It's late. |
| T: We'll be on time. | W: We wait for Sydney. | J: John will be joining us. | M: Mary will come. |

1. A & C
2. M ↔ ~L
3. ~L → W
4. (T v L) & ~(T & L)
5. E → C
6. (W & M) & (W → ~T)
7. J → (M & ~E)
8. C v A
9. C & [A & {W → (L → ~T)}]
10. L → (J & M)
11. W → E
12. (T v ~W) & L
13. J ↔ (M & ~E)
14. (E v A) v M
15. ~(W & J) → C
16. (C & T) ↔ ~J
17. ~(W → T)
18. (L v T) & ~(L & T)
19. (C → A) v ~A
20. T v L

> **THINKING AHEAD**
>
> Think about any statement from Exercise 7b and all the possible trues and falses you could attach to the atomic statements that make it up. Imagine that you have to construct a system that can represent what the overall truth status is of that statement in each of those cases.

## INTRODUCING TRUTH TABLES

| A | ~A |
|---|---|
| T | F |
| F | T |

A connective's function is the way in which it generates a truth status for the proposition, based on the truth-values of the component parts. In Unit One we characterized connectives as expressing certain English language terms:

~ not
& and
v or
→ if ... then
↔ if and only if

Now we will *define* the connectives in terms of their truth functions.

Our logic system is called a bivalent logic system, which means that every statement has to be *either* true or false, and it cannot be both at the same time. This is called the *truth-value* of the proposition. In order to understand the function of a connective, we therefore need a way to represent all the component parts of a proposition, and all the possible combinations of truth values for its component parts, and to show the resulting truth-values of their combinations. The truth table is an excellent way to do this.

Creating a truth table to analyze a statement requires the following steps. Once you've symbolized the statement with its connectives and using only capital letters for your translation of the component sentence(s), you can begin.

1. Temporarily ignore all the connectives. You'll be left with all the different capital letters by themselves.
2. Count these atomic statements, giving you some number, $n$.
3. Create a table that has at the top enough columns for each of the atomic statement letters to have their own box and a separate box to the far right for the original statement you're analyzing.
4. Underneath this top row complete the table by adding $2^n$ number of rows.
5. Place the letters representing the atomic statements starting in the upper left box and working your way across the top row.
6. Take the complete statement that you want to analyze and place it in the box to the right of the last of your statement letters.
7. You now begin putting Ts and Fs underneath the atomic statements so that you can cover *all possible combinations of true and false*. Start with the letter in the box right next to the original statement. Begin with T and then write F in the box below that, and continue alternating each box underneath the statement letter until you're done with that column.

8. Move to the next column under the next statement letter and work your way down the column filling the boxes in, this time putting *two* Ts, followed by *two* Fs, followed by *two* Ts, followed by *two* Fs, until you're at the bottom.
9. Move to the next column and repeat the process, only this time with *four* boxes of Ts, followed by *four* boxes of Fs, and so on.
10. Repeat the process, doubling the number each time until you reach the first column and you've filled in all the boxes. You have now represented every possible truth-value combination for the single statements.

We'll demonstrate this much of the process by using the sentence,

1. It's not true that George Washington was born in Thailand.

Translating it with **LOLA**-Lite, we get ~G. Begin by breaking down the statement into its components, the atomic statements, and the connectives. Ignore the connectives for now, and look at the remaining letter(s), and all that's left is "G." So now we can move on and create a table with enough columns for each of the individual statement letters and the original statement to have their own individual boxes. This gives us 2 columns.

Then we add $2^n$ number of rows underneath this. We have only one single statement, so that gives us 2 to the first power, which is, of course, just 2. So, we'll create two rows underneath this.

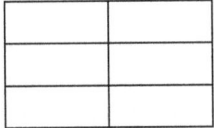

We then place the statement letters in the upper left box and work our way across until we've put them all in boxes, and then we put the original statement in the remaining box.

| G | ~G |
|---|----|
|   |    |
|   |    |

We are now ready to assign truth-values for all the single statements.

| G | ~G |
|---|---|
| T |   |
| F |   |

We're now ready to demonstrate all the possible truth values for "~G."

## USING TRUTH TABLES TO DEMONSTRATE A CONNECTIVE'S FUNCTION: BASIC TRUTH TABLES

Truth tables will always follow the same steps we just laid out above. However, having constructed the first part of the truth table, we now need to specify the function of the connective, in order to complete the truth table. In what follows we'll lay out the rules that cover each of the five connectives.

*Negation (~)*

Since the tilde negates whatever it is operating on, it can be seen as *reversing* the truth-value of the positive statement. So, if the value of a statement was listed as true, adding a tilde to it would create a *new* truth value, namely it would make it false. If the statement was false, then by negating it the tilde makes the proposition into a true one.

This gives us:

| G | ~G |
|---|---|
| T | F |
| F | T |

Note that the F and the T in the right column goes directly under the connective. Our example above is fairly straightforward. If the atomic statement "G" is true, then applying a tilde to it would make it false. If "G" is false, then applying a tilde to it says that it's not true that "G" is false, and hence, it must be true.

*Conjunctions (&)*

2. Bob and Dave are going to the movies.

This one sentence is actually two statements, "Bob is going to the movies" and "Dave is going to the movies" joined by a conjunction. The two statements that make up this conjunction are called the *conjuncts*.

Begin by symbolizing each of them with different capital letters:

B: Bob is going to the movies.
D: Dave is going to the movies.

Then symbolize the overall statement:

**B & D**

Now, using the steps we listed above, start a truth table.

| B | D | B & D |
|---|---|-------|
| T | T |       |
| T | F |       |
| F | T |       |
| F | F |       |

In order to complete the truth table, we need the rule for how a conjunction functions.

A conjunction is essentially the simultaneous assertion of two statements; it is saying that *both* these statements are true. Any situation in which *either* of the two statements is false would falsify the conjunction as a whole. Another way of saying this would be that conjunctions are true *only* when both the component parts are true. That would give us the following table:

| B | D | B & D |
|---|---|-------|
| T | T | T     |
| T | F | F     |
| F | T | F     |
| F | F | F     |

Notice that the only time a conjunction is true is if *both* conjuncts are true.

*Disjunctions (v)*

The next connective we'll consider is the disjunction. Take the following statement,

3. Either we'll go to the party or the beach.

Symbolize the component statements with capital letters:

B: We'll go to the beach.
P: We'll go to the party.

The resulting compound statement is:

**B v P**

Set up the truth table:

| B | P | B v P |
|---|---|-------|
| T | T |       |
| T | F |       |
| F | T |       |
| F | F |       |

In Chapter Five we explained that the default interpretation of a disjunction is that either one proposition or the other or both of them could be true. So, the *only* time a disjunction is *false* is when *both* disjuncts are false. This gives us:

| B | P | B v P |
|---|---|-------|
| T | T | T     |
| T | F | T     |
| F | T | T     |
| F | F | F     |

> **THINKING AHEAD**
>
> Think about how you would evaluate a biconditional statement with one statement false and one true? What would be the truth value of the statement as a whole? What about a biconditional statement with two false statements? How would you evaluate a conditional in which the antecedent was true and the consequent false? How about a statement in which both the antecedent and consequent are false?

*Biconditionals (↔)*

The next example is the biconditional or double arrow. Use the following sentence:

4. The chemical reaction will work if and only if our equations are correct.

Symbolizing the statements gives us:

C: The chemical reaction will work; E: Our equations are correct.

The compound statement is:

$$C \leftrightarrow E$$

The basic idea behind the double arrow is that the biconditional is true when, and only when, the truth-values of the two sides are the same. So, if the left side is true and the right side is true, then the statement as a whole is true. If the right side is false and the left side is *also* false, then the statement as a whole is true. This means that whenever the truth-values are the *same* for the two sides, the statement as whole is *true*. When they *differ*, the statement as a whole is *false*.

Hence, when we set up the truth table, we have,

| C | E | C ↔ E |
|---|---|---|
| T | T | T |
| T | F | F |
| F | T | F |
| F | F | T |

*Conditionals (→)*

The last of the connectives is the arrow. Because conditionals are especially significant in everyday life, and in logic, and because the nature of logical conditionals sometimes throws people, we'll examine it more slowly. Remember that conditionals have two parts to them, the *antecedent* (the part that comes before the arrow) and the *consequent* (the part that comes after the arrow). It is critical that you be able to quickly and clearly identify which part of the conditional is the antecedent and which part is the consequent.

Remember that what a conditional says is that *if* the antecedent is true, *then* the consequent will be true. *Nothing more or nothing less than that*. In order to think more clearly about the truth functional nature of the arrow, read sentence 5 and ask yourself under what conditions would you say that Bob had broken his promise?

5. (Bob promised that) **if Cindy keeps her room clean, he'd get her a puppy**.

Symbolizing Bob's statement gives us:

R: Cindy keeps her room clean; P: Bob gets Cindy a puppy. The compound statement is then symbolized: **R → P**

Set up the truth table,

| R | P | R → P |
|---|---|-------|
| T | T |       |
| T | F |       |
| F | T |       |
| F | F |       |

Begin by making sure you can identify the antecedent and the consequent. The R is the antecedent, and the P is the consequent. As we suggested above, rather than think of this in terms of bare truth or falsehood, think of it in terms of a promise. Under what conditions would we say that Bob had broken his promise? Under what conditions did Bob not *speak the truth*?

Let's begin with the bottom row.

If it turns out that Cindy *didn't* clean her room, then would we claim that Bob *lied* if he *didn't* get Cindy a puppy? No. His only obligation to her is to follow through on the puppy *if she cleaned her room*, and because R reads false in that row, it means she didn't clean her room. So, if he doesn't buy her the puppy, that is if P also reads as false, the sentence as a whole doesn't constitute a case of Bob breaking his promise. He hasn't lied; he hasn't stated a falsehood. Since we have a bivalent system of logic and all propositions *have* to be either true or false, and we've decided that this isn't a case of Bob speaking falsely, we put down a T in the last box. Hence, whenever you have a conditional and *both the antecedent and the consequent are false, the statement as a whole is true.*

| R | P | R → P |
|---|---|-------|
| T | T |       |
| T | F |       |
| F | T |       |
| F | F | T     |

Now we go to row 2 and we have a scenario in which Cindy does clean her room and yet Bob *doesn't* buy her a puppy. Would we say that this is a case of lying or breaking a promise? Yes. So, when you have a situation in which the *antecedent is true, but the consequent is false, then you have a falsehood.*

| R | P | R → P |
|---|---|-------|
| T | T |       |
| T | F | F     |
| F | T |       |
| F | F | T     |

We now turn to the third row in which Cindy didn't clean her room, but good ole Bob got her the puppy anyway. As in the case of the bottom row, there isn't any lie or broken promise here. Bob never said what he would or wouldn't do in a case in which she *didn't* clean her room. As we said before, his only obligation was to get her the puppy if she cleaned her room. Once again, in a bivalent system, if a statement is not false that means it has to be true. Therefore, conditionals in which the *antecedent is false and the consequent is true will be true.*

| R | P | R → P |
|---|---|-------|
| T | T |       |
| T | F | F     |
| F | T | T     |
| F | F | T     |

Finally, row 1, we have the scenario in which Cindy cleans her room and Bob gets her the puppy. Bob kept his promise, so we list this as true. *Whenever both the antecedent and consequent of a conditional are true then the conditional as a whole is true.*

| R | P | R → P |
|---|---|---|
| T | T | T |
| T | F | F |
| F | T | T |
| F | F | T |

So, for conditionals, the only time a conditional statement is false is when the antecedent is true and the consequent is false.

This concludes the basic truth tables. Essentially, these tables for each of the connectives in our system visually *define the truth-functionality of each of the connectives.*

| A | ~A |
|---|---|
| T | F |
| F | T |

| A | B | A v B |
|---|---|---|
| T | T | T |
| T | F | T |
| F | T | T |
| F | F | F |

| A | B | A ↔ B |
|---|---|---|
| T | T | T |
| T | F | F |
| F | T | F |
| F | F | T |

| A | B | A → B |
|---|---|---|
| T | T | T |
| T | F | F |
| F | T | T |
| F | F | T |

| A | B | A & B |
|---|---|---|
| T | T | T |
| T | F | F |
| F | T | F |
| F | F | F |

We can now use these basic truth tables to analyze the truth values of more complex statements.

## MORE COMPLEX STATEMENTS AND TRUTH TABLES

As was the case in the previous section, our primary goal is understanding the nature of connectives, and so we'll continue to use LOLA-Lite. To make this easier, we'll supply the translation scheme immediately after the sentence rather than supply a separate, detailed translation scheme.

**Example Set 7a**
1. Assuming Bob goes to town, then Frank will stay home and watch Cindy. (B: Bob goes to town; F: Frank stays home; W: Frank watches Cindy.)
2. Either Cindy or Lou will win the game, but not both of them. (C: Cindy wins the game; L: Lou wins the game.)
3. Either Sam gets a loan and buys the house or, if he can't afford that, he moves in with his Mom. (L: Sam gets a loan; B: Sam buys a house; A: Sam can afford the house; M: Sam moves in with his Mom.)
4. It's not dangerous if and only if you pay attention and know what you're doing. (D: It's dangerous; P: You pay attention; K: You know what you're doing.)
5. The test will be on Monday only if we get through this chapter or you don't annoy me. (T: The test is on Monday; C: We get through this chapter; A: You annoy me.)

All of these compound statements have constituent parts that are themselves compound statements. The same basic techniques we developed in the first six chapters also apply here. You begin by constructing a translation scheme. Then, if you have multiple operators, you identify all the operators in the sentence. Next, you identify which one is the *main* operator and then group all the parts of the statement with parentheses so that the statement's structure is clear.

However, what we said earlier applies here as well. While it usually isn't difficult to spot the operators there is no mechanical approach that you can simply apply to ascertain which of the operators is the *main* one. The goal is to symbolically represent the sentence as accurately as possible. In sentence 1, for example, there is a conditional and a conjunction. Ask yourself whether the fundamental thrust of the sentence is an assertion that two propositions are the case, one of which happens to be a conditional or is it asserting a conditional relationship between two propositions, one of which is a conjunction. Once you look at it this way, it should be clear that it is the latter.

1. B → (F & W)

Sentence 2, which is an example of the exclusive sense of "or," has a disjunction, a conjunction, and a tilde. In this case, the main operator is the conjunction, which joins two propositions, one of which is a disjunction and the other of which is itself a compound statement composed of a tilde and a conjunction. This gives you,

**2. (C v L) & ~( C & L)**

Sentence 3 is a disjunction with a conjunction on one side and a conditional with a negated antecedent on the other side resulting in,

**3. (L & B) v (~A → M)**

Sentence 4 is a biconditional with a negation on one side, and a conjunction on the other, giving you,

**4. ~D ↔ (P & K)**

Sentence 5 is a conditional with a disjunction as its consequent giving you,

**5. T → (C v ~A)**

Having symbolized all five sentences we can now construct truth tables for them.

Begin by counting the number of atomic statements in sentence 1 and then construct the appropriate sized truth table. Since there are 3 atomic statements, that means you'll have four columns (one for each of the single statements and one for the compound statement as a whole), and one row at the top with $2^3$ (= 8) *truth-input* rows underneath it.

| B | F | W | B → (F & W) |
|---|---|---|---|
| T | T | T | |
| T | T | F | |
| T | F | T | |
| T | F | F | |
| F | T | T | |
| F | T | F | |
| F | F | T | |
| F | F | F | |

Because there is more than one operator in the statement, you'll need to evaluate the truth-value of the constituent parts first, and then when you've done them all you'll be able to evaluate the truth-value of the statement as a whole. Therefore, *always begin by analyzing the connective with the smallest scope* (F & W, in this case) as though it were all by itself. *Remember that for a conjunction, both conjuncts have to be T for the statement to read T.* Write the truth value for the conjunction directly under the &.

| B | F | W | B → (F & W) |
|---|---|---|---|
| T | T | T | T |
| T | T | F | F |
| T | F | T | F |
| T | F | F | F |
| F | T | T | T |
| F | T | F | F |
| F | F | T | F |
| F | F | F | F |

You have now established the truth values of the *consequent* of the conditional, B → (F & W). Now consider the antecedent. Although we already have the values for B in the first column, *it often helps you keep track if you rewrite the truth-values of single statements, so you can just read across the compound*. We've done this below for B.

| B | F | W | B → | (F & W) |
|---|---|---|---|---|
| T | T | T | T | T |
| T | T | F | T | F |
| T | F | T | T | F |
| T | F | F | T | F |
| F | T | T | F | T |
| F | T | F | F | F |
| F | F | T | F | F |
| F | F | F | F | F |

Our table is now to the point where we can demonstrate all the possible truth-values for the statement as a whole. Since our statement is a conditional, we know that *the only time it will be false is when the antecedent is true and the consequent is false.*

Again, for each row, we can just read off from what we've written down: if that row "goes from" T to F, the value of the conditional for that row is F, otherwise it is true. We've finished the truth table below, putting the truth status of the conditional in bold and italic so that you can see it more clearly.

| B | F | W | B → (F & W) | |
|---|---|---|---|---|
| T | T | T | T ***T*** | T |
| T | T | F | T ***F*** | F |
| T | F | T | T ***F*** | F |
| T | F | F | T ***F*** | F |
| F | T | T | F ***T*** | T |
| F | T | F | F ***T*** | F |
| F | F | T | F ***T*** | F |
| F | F | F | F ***T*** | F |

By reading the truth table you can see that there are three situations in which the original statement is false. What's more, you can actually read off what the conditions are that would make the original statement false.

Row 2 says the statement is false *if*
    a) Bob goes to town.
    b) Frank stays at home.
    c) Frank neglects to watch Cindy.

Row 3 says the statement is false *if*
    a) Bob goes to town.
    b) Frank doesn't stay home.
    c) Frank watches Cindy.

Row 4 says the statement is false *if*
    a) Bob goes to town.
    b) Frank doesn't stay home.
    c) Frank doesn't watch Cindy.

All other combinations of trues and falses would make the statement as whole true.

Even though sentence 1 wasn't particularly complicated, it involved constructing a much larger truth table than any we did earlier in this chapter. *Keep in mind that it isn't the complexity of the statement that determines the size of the truth table, it's only the number of single statements.* For example, even though sentence 2 is longer, it has only two single statements so it will have only four rows beneath the top row.

| C | L | (C v L) & ~(C & L) |
|---|---|---|
| T | T |   |
| T | F |   |
| F | T |   |
| F | F |   |

In this statement, there are four operators at work: the conjunction of C and L, the negation of that conjunction, the disjunction of C and L, and then the conjunction functioning as the main connective. How to begin?

This requires some care since you are faced with multiple operators, and what is more, some of these connectives are dependent upon the resolution of other connectives for their truth status. As we said above, *work from connectives with the smallest scope to connectives with the widest scope.* Another way to think about this is to work from the "inside out." Look at the formula and identify those cases where one connective, what we'll call the inner connective, is being operated on by *another* connective, what we'll call the outer connective. That's exactly what's happening in the case of ~(C & L). You have an inner formula, the conjunction (C & L), and then on the outside of it, you have a tilde, saying that this *conjunction* is false. You have to evaluate the conjunction, which is the inner connective, first, and then you can evaluate the outer connective, the tilde.

| C | L | (C v L) & ~(C & L) |
|---|---|---|
| T | T | T |
| T | F | F |
| F | T | F |
| F | F | F |

Now the tilde will reverse all the truth values of that conjunction. (One way to keep track of what you've done is to cross out each of the inner truth values after you've moved onto the outer function.)

| C | L | (C v L) & ~(C & L) |
|---|---|---|
| T | T | F ~~T~~ |
| T | F | T ~~F~~ |
| F | T | T ~~F~~ |
| F | F | T ~~F~~ |

This means you've completed the right side of the conjunction and you can move onto the left side. Here all you have to do is determine the truth values for the disjunction. A disjunction is only false when both disjuncts are false:

| C | L | (C v L) & ~ (C & L) |
|---|---|---|
| T | T | T   F ~~T~~ |
| T | F | T   T ~~F~~ |
| F | T | T   T ~~F~~ |
| F | F | F   T ~~F~~ |

You've now completed the truth values for both sides of the conjunction and you can now read the conjunction as a whole. Again, it can be helpful to cross out the truth values of the two sides as you put down the truth value for the formula as a whole.

| C | L | (C v L) & ~ (C & L) |
|---|---|---|
| T | T | ~~T~~ **F** ~~F~~ ~~T~~ |
| T | F | ~~T~~ **T** ~~T~~ ~~F~~ |
| F | T | ~~T~~ **T** ~~T~~ ~~F~~ |
| F | F | ~~F~~ **F** ~~T~~ ~~F~~ |

Sentence 3 illustrates one of the potential problems for the method of truth tables. Because there are four single statements, we need to construct a truth table that has one row on the top and *sixteen* rows underneath it.

| A | B | L | M | (L & B) v (~A → M) |
|---|---|---|---|---|
| T | T | T | T | |
| T | T | T | F | |
| T | T | F | T | |
| T | T | F | F | |
| T | F | T | T | |
| T | F | T | F | |
| T | F | F | T | |
| T | F | F | F | |
| F | T | T | T | |
| F | T | T | F | |
| F | T | F | T | |
| F | T | F | F | |
| F | F | T | T | |
| F | F | T | F | |
| F | F | F | T | |
| F | F | F | F | |

Remember, though, even though this table is longer, the initial set up doesn't change: The innermost column alternates Ts and Fs, and each column working out just doubles the consecutive Ts and Fs, till you reach the outermost column. But now we have to fill it out and read it. Don't be intimidated by the task, we still just go one step at a time, just as we did for the smaller tables. The key to larger tables, really, is to continue working methodically and *neatly*. Once again, we'll find the connective(s) with the smallest scope. The compound is a disjunction, so you can really begin on either side. We'll start with the left-hand disjunct, since it is a conjunction, and therefore it's easy to fill out its values.

We've highlighted the columns you need to fill out values for the conjunction—this is just another "keeping track" method. Below is the completed column for the left-hand disjunct.

| A | B | L | M | (L & B) v (~A → M) |
|---|---|---|---|---|
| T | T | T | T | T |
| T | T | T | F | T |
| T | T | F | T | F |
| T | T | F | F | F |
| T | F | T | T | F |
| T | F | T | F | F |
| T | F | F | T | F |
| T | F | F | F | F |
| F | T | T | T | T |
| F | T | T | F | T |
| F | T | F | T | F |
| F | T | F | F | F |
| F | F | T | T | F |
| F | F | T | F | F |
| F | F | F | T | F |
| F | F | F | F | F |

# CHAPTER SEVEN: CONNECTIVES AND TRUTH TABLES

Now, the right side: This is just a conditional, so we know that it is only false if the antecedent is true and the consequent is false. *Be sure to write down values for ~A before evaluating the entire conditional.* The truth values for the right disjunct, then, are:

| A | B | L | M | (L & B) v (~A → M) |   |   |
|---|---|---|---|---|---|---|
| T | T | T | T | T | F | T |
| T | T | T | F | T | F | T |
| T | T | F | T | F | F | T |
| T | T | F | F | F | F | T |
| T | F | T | T | F | F | T |
| T | F | T | F | F | F | T |
| T | F | F | T | F | F | T |
| T | F | F | F | F | F | T |
| F | T | T | T | T | T | T |
| F | T | T | F | T | T | F |
| F | T | F | T | F | T | T |
| F | T | F | F | F | T | F |
| F | F | T | T | F | T | T |
| F | F | T | F | F | T | F |
| F | F | F | T | F | T | T |
| F | F | F | F | F | T | F |

All that's left is to determine the truth-values for the disjunction as a whole. The values of the disjunction are determined by the values of each disjunct. As we finish the table below, we've put in **bold** the values of each disjunct, and for the values of the whole disjunction we've put in ***bold italics***.

| A | B | L | M | (L & B) v (~A → M) | | | |
|---|---|---|---|---|---|---|---|
| T | T | T | T | **T** | ***T*** | F | **T** |
| T | T | T | F | **T** | ***T*** | F | **T** |
| T | T | F | T | **F** | ***T*** | F | **T** |
| T | T | F | F | **F** | ***T*** | F | **T** |
| T | F | T | T | **F** | ***T*** | F | **T** |
| T | F | T | F | **F** | ***T*** | F | **T** |
| T | F | F | T | **F** | ***T*** | F | **T** |
| T | F | F | F | **F** | ***T*** | F | **T** |
| F | T | T | T | **T** | ***T*** | T | **T** |
| F | T | T | F | **T** | ***T*** | T | **F** |
| F | T | F | T | **F** | ***T*** | T | **T** |
| F | T | F | F | **F** | ***F*** | T | **F** |
| F | F | T | T | **F** | ***T*** | T | **T** |
| F | F | T | F | **F** | ***F*** | T | **F** |
| F | F | F | T | **F** | ***T*** | T | **T** |
| F | F | F | F | **F** | ***F*** | T | **F** |

# CHAPTER SEVEN: CONNECTIVES AND TRUTH TABLES

> **PRACTICE**
> 
> Before you continue, set up and complete truth tables for the remaining two sentences. After doing these, check the following completed tables for these sentences.

Sentence 4: ~D ⟷ (P & K)

| D | K | P | ~D ⟷ (P & K) |
|---|---|---|---|
| T | T | T | ~~F~~   F   ~~T T T~~ |
| T | T | F | ~~F~~   T   ~~F F T~~ |
| T | F | T | ~~F~~   T   ~~T F F~~ |
| T | F | F | ~~F~~   T   ~~F F F~~ |
| F | T | T | ~~T~~   T   ~~T T T~~ |
| F | T | F | ~~T~~   F   ~~F F T~~ |
| F | F | T | ~~T~~   F   ~~T F F~~ |
| F | F | F | ~~T~~   F   ~~F F F~~ |

Sentence 5: T → (C v ~A)

| A | C | T | T → (C v ~A) |
|---|---|---|---|
| T | T | T | ~~T~~ T   ~~T T F~~ |
| T | T | F | ~~F~~ T   ~~T T F~~ |
| T | F | T | ~~T~~ F   ~~F F F~~ |
| T | F | F | ~~F~~ T   ~~F F F~~ |
| F | T | T | ~~T~~ T   ~~T T T~~ |
| F | T | F | ~~F~~ T   ~~T T T~~ |
| F | F | T | ~~T~~ T   ~~F T T~~ |
| F | F | F | ~~F~~ T   ~~F T T~~ |

## Exercise 7c

Set up and complete truth tables for the following LOLA-lite sentences.

1. A & (~B v C)
2. C ⟷ (D & C)
3. D v (K → L)
4. ~M v ~N
5. ~M v (~M v ~N)
6. (P & Q) v ~(P & Q)
7. (M v ~M) → D
8. (D v ~D) → (M & ~M)
9. C → (L → Z)
10. K v (K → ~K)

Symbolize the following sentences and then construct truth tables for them.

11. People either like Professor Marshall's economics class or they don't. (E: People like Professor Marshall's economics class.)
12. If the hurricane hits the coast, there're going to have massive winds. (H: The hurricane hits the coast; W: There will be massive winds.)
13. Either Ryan picks up his sister from classes and takes her to her lessons or his dad will revoke Ryan's car privileges and he can walk everywhere. (P: Ryan picks up his sister from classes; T: Ryan takes his sister to her lessons; R: Ryan has his car privileges revoked; W: Ryan can walk everywhere.)
14. Not everything good in life is free. (F: Everything good in life is free.)
15. We'll cross that bridge when we come to it just in case there is a bridge. (C: We cross the bridge when we come to it; B: There is a bridge.)
16. Unless Cyril doesn't think it's a good idea, Peter is planning on moving to New York this summer. (G: Cyril thinks it's a good idea; N: Peter plans on moving to New York this summer.)
17. There's a problem with the carburetor, although if Morton has to spend much money on it, he'll just scrap the car and buy a used one. (C: There's a problem with the carburetor; M: Morton has to spend much money on the car; S: Morton scraps the car; U: Morton buys a used car.)
18. The lamps go on automatically at dusk, and then, if there's not enough light to see, you can turn on the rest of them manually. (A: The lamps go on automatically; E: There's enough light to see; T: A person can turn on the rest of the lights manually.)
19. The latest news is not good; however, there's always hope and prayer. (N: The news is good; H: There is always hope; P: There is always prayer.)
20. You can always lead a horse to water, but you can't make him drink. (L: You lead a horse to water; D: You make the horse drink.)

---

**KEEP IN PRACTICE**

Give a complete translation (names, quantifiers, predicates, etc.) of each sentence above.

## KINDS OF STATEMENTS: CONTINGENT STATEMENTS, LOGICAL TRUTHS, AND LOGICAL FALSEHOODS

Every statement in LOLA is either true or false *on some particular interpretation*. In LOLA an interpretation simply means an assignment of either truth or falsehood to all relevant atomic statements. What we mean by this is that we don't investigate whether those statements actually match up with the real world. We don't even go into a discussion of the methods of how you might effectively conduct such an investigation. Critical thinking courses often address such important questions, as do theory and methods courses in various disciplines. But in our symbolic logic we simply *assign* each atomic statement a status of T or F. This may seem that we are ignoring something of extreme importance, namely whether statements are "really" true. Working with "truth assignments" and LOLA, though, lets us discover important truths *about* statements that go beyond any individual statement.

Think of the statement from Example Set 7a: The test will be on Monday only if we get through this chapter or you don't annoy me. Our symbolic representation of this statement was: **T → (C v ~A)**. When we ran this truth table, we determined that this statement is false under only one interpretation: When T is true, C is false and A is true. But the truth table tells us even more than this: **it tells us under what conditions any statement with that same form will be true or false**. For instance:

> If cats could fly, then birds would be nervous unless they couldn't swim.
> Proposition A will pass only if Democrats show up in force or Republicans don't make it an issue.
> Truckers won't agree to the contract unless either their wage demands are met or management doesn't try to cut their benefits again.

All of these statements (and countless others) have the same form as T → (C v ~A): a conditional whose antecedent is a single statement and whose consequent is a disjunction, with one disjunct being another single statement and the other disjunct being the negation of yet another single statement. The point is that for any given compound statement, the *truth table gives all possible truth value conditions for that statement and any statement of the same form*.

Truth tables also show us that compound statements must have one of only three possible "truth statuses." The *final column* (i.e., the column under the main connective) of a truth table must show either all Ts, all Fs, or some Ts and some Fs. Any statement that can be true on one interpretation and false on another is called a *contingent* statement. By their very nature, all single statements are contingent statements, since we defined a statement as a sentence that can be true or false. Compound statements that are true under any interpretation are called *logical truths*—we say, for short, that they're L-true. (They're also sometimes called tautologies.) Compound statements that are false under any interpretation are called *logical falsehoods*—we say, for short, that they're L-false. (They're also sometimes called self-contradictions.) We can easily present examples of all three kinds of statements through the method of truth tables.

> It's raining.
> It's raining **or** it's not raining.
> It's raining **and** it's not raining.

Take the following statement,

Either it's raining or it's not raining. (R: It is raining.)

Symbolize it:

**R v ~R**

Now set up a truth table for it.

| R | R v ~R |
|---|--------|
| T |        |
| F |        |

Now, reading from left to right interpret the statement in light of the truth-values of the single statements.

The first row, where the single statement R is listed as T gives us an interpretation where the statement R v ~ R comes out as true.

| R | R v ~R |
|---|--------|
| T | T T F  |
| F |        |

The second row, in which the single statement R is listed as F also gives us an interpretation in which the statement R v ~R comes out as true.

| R | R v ~R |
|---|--------|
| T | T T F  |
| F | F T F  |

So it doesn't matter what interpretation we construct, R v ~R will always come out as true. R v ~R is *true under every possible interpretation*. What that means is that whatever the facts are—whether it's really raining or not—this statement is true. So we don't have to know anything about the facts to know that this statement must be true. We don't even have to know what the symbols of a statement of this form stand for: *any statement that has this pattern will always be true.* Thus, even though we don't tell you what English sentence G stands for, you know already that G v ~G is true.

Our next kinds of statement, logical falsehoods, are those statements that are *false under every possible interpretation*. For example, the following statement:

It is hot outside and it's not hot outside. (H: It's hot outside.)

Symbolize it:

**H & ~H**

Set up the truth table and read the first row, in which H reads T. This gives us an interpretation of H & ~ H that comes out false.

| H | H & ~ H |
|---|---------|
| T | T̶ F F̶  |
| F |         |

The second row, in which H reads F gives us an interpretation of H & ~H that also comes out as false.

| H | H & ~ H |
|---|---------|
| T | T̶ F F̶  |
| F | F̶ F T̶  |

So the statement H & ~H comes out as false under every possible interpretation. Again, if we can see the structure of this statement, we don't even have to know what the symbol "H" stands for; it is an *intrinsic property of statements of this form or pattern that they will always be false*. So G & ~G must also be false.

Finally, a statement is contingent if there is some interpretation (one or more) in which it comes out false and some interpretation (one or more) in which it comes out as true. Take a simple conditional statement: If Paul wants to go to law school, he needs to take the LSATs. (P: Paul wants to go to law school; N: Paul needs to take the LSATs.) Symbolize it:

$$P \rightarrow N$$

Set up and complete the truth table.

| P | N | P → N |
|---|---|---|
| T | T | ~~T~~ T ~~T~~ |
| T | F | ~~T~~ F ~~F~~ |
| F | T | ~~F~~ T ~~T~~ |
| F | F | ~~F~~ T ~~F~~ |

When you fill in all the rows you find that at least one row yields an interpretation of T for the statement as a whole and at least one row yields an interpretation of F. This means the statement is contingent. You've seen contingent statements before: look back at the basic truth tables; you'll notice that each of these is contingent.

Before we leave this topic, there's something that it's quite important to understand: the difference between these three categories, and ordinary truth and falsehood. One of the rows in the truth table for R v ~R above (one of its interpretations) represents the real world right here and now (depending on whether it's raining right here and now or not). But, as we've seen, whichever row represents the real world, R v ~R is true: the facts don't matter. Similarly, one of the rows for H & ~H represents the real world, but that statement is false whatever the facts about the temperature outside. Is P → N in fact true or false? That depends on the facts. If it's true that Paul wants to go to law school, but false that Paul needs to take the LSATs, then the real world is represented by the second row of the truth table, and P → N is in fact false. *But it isn't L-false.* To sum up: every L-true sentence is also in fact true; every L-false sentence is also in fact false; some contingent sentences are in fact true, and others are in fact false.

## Exercise 7d

Set up and complete truth tables for the following LOLA-Lite sentences, and state whether each statement is L-true, L-false, or contingent.

1. (D & E) v ~(D & E)
2. (G v ~G) → (B & ~B)
3. (D v K) & (K → L)
4. (P → W) ↔ (~P v W)
5. ~(E & ~F) ↔ ~(E → F)
6. [M → (A v T)] & (M & A)
7. ~[(Q → S) & ~Q]
8. (~R → S) & ~(~S → R)
9. (O ↔ C) ↔ [(O & C) v (~O & ~C)]
10. [(I & J) v (I v J)] v (I → J)

## Exercise 7e (Translation Review)

For the following statements, give complete LOLA translations (*not* LOLA-Lite) according to the given translation scheme.

### TRANSLATION SCHEME
RD: Everything

| e: Wally | l: Lane | c: Carol | m: Mark |
| --- | --- | --- | --- |
| p: Pam | Px: x is a person | Ax: x is an aardvark | Vx: x is a singer |
| Wx: x is a whale | Fx: x is a fish | Qx: x quits | Cx: x is a car |
| Sx: x is speedy | Fxy: x is faster than y | Lxy: x likes singing to y | Kxy: x likes y |

1. Wally is speedy, but somebody is faster.
2. Mark and Carol quit, but not everyone will.
3. An aardvark is not a car nor is it a person, nor a whale.
4. Lane is not a singer, but everyone likes singing to Pam.
5. A whale is neither a fish nor a car, but then neither is Carol.
6. People are not faster than cars, but whales are if they don't quit.
7. Some whales like singing to some whales, but not all whales are singers.
8. Wally is speedy and he doesn't quit, but most people do.
9. If an aardvark is faster than a fish, then it is speedy.
10. People like speedy cars.
11. A person is a singer just in case they like singing to somebody.
12. A person quits being a singer only if they don't like singing to somebody.
13. Speedy fish like singing to Pam.
14. Not everyone quits, although some people do.
15. Lane and Carol are faster than every aardvark, but not any whales.

16. Mark is a whale, not a person, though he likes singing both to whales and people.
17. If it's a car, then it's faster than either Mark or Pam or Wally or any person for that matter.
18. Not everything is either a fish or a person or a whale.
19. It's a singer only if it's a person and not a fish.
20. Everything likes singing to something.

# Chapter Eight: Truth Trees

**What's Up?**
Truth Trees
Open and Closed Branches
L-true, L-false, Contingent

In the first seven chapters we've analyzed statements. We've looked at many different kinds of statements, learned how to symbolize them to best represent their logical structure, introduced truth tables, seen how tables can express the truth functional nature of connectives, and learned the difference between contingent, logical truths, and logical falsehoods. In this chapter, we'll continue with our study of the truth functional nature of connectives and the truth categories of L-truth, L-falsehood, and contingency by introducing a new technique called truth trees.

### WHY TREES?

Truth tables are an excellent means of representing certain aspects of statements, such as the way connectives function. What's more, they have the advantage of being mechanical. You simply set the truth table up for a given task, plug in the formula, and then read off the rows to determine the truth-value of a statement. We'll later also use the method of tables to evaluate relationships between two statements and also test for argument validity. Despite the usefulness of truth tables, however, we will introduce you to a new technique called truth trees. We have at least three reasons for this.

First, knowing a variety of techniques for analysis within logic will always be beneficial. The more different ways you can go at a problem, the greater your ability to analyze another person's ideas and to clarify your own. Because of this we include sections on truth tables, truth trees, and the method of proof. All of these techniques have their strengths and limitations, and each technique captures an aspect of the way we think.

Second, truth tables can quickly become unwieldy if there are more than three atomic statements. Writing out a sixteen-row table, and being able to clearly read off the truth assignments, can be taxing. Imagine trying to work with a compound statement that includes five or six atomic statements. There are ways to deal with this, but often at the cost of sacrificing the mechanical features of the truth table.

The third problem is related to the second and is perhaps more consequential. Truth tables are not able to handle predicates, quantifiers, and identity statements. That means that if we wanted to use tables to evaluate the truth status of a statement we have to symbolize it using only the approach we introduced in Chapter Seven, LOLA-Lite. Each statement would have to be symbolized as simply a capital letter, thereby leaving out all the information that is normally carried in a statement. As we said in Chapter Seven, on those occasions, in which the task you're facing doesn't require a more sophisticated representation of the statements involved, truth tables may be adequate. However, many times you'll need to be able to represent the statement in as much depth as possible, and so being restricted in your method of symbolization will be a severe drawback.

The basic idea behind a truth tree relies on what we said in the previous chapter: statements in LOLA are either true or false *under some interpretation*. So, if you have a conditional, and the antecedent is designated as T and the consequent is designated as F, then the conditional as a whole is false. If you have a different interpretation in which the antecedent is designated as T and the consequent is designated as T, then the conditional as a whole is true. What a tree does is to trace out the various possible interpretations of one or more statements. Trees can be done whether you're working with predicates, quantifiers, individuals or the more simplified LOLA-lite approach introduced in the previous chapter.

While both truth tables and truth trees rely on the relationships defined in the basic truth tables, we get at the information we want in different ways. In a truth table, we "build up." Starting from the connective with the smallest scope, we work to the main connective and, in the final column, we see all the possible truth values for the compound statement. On a truth tree, we "break down." Starting from the original statement, we ask at each step, "Under what conditions, or interpretations, would this statement be TRUE?" Then, we write down those conditions in a systematic way. At each step, then, we only indicate when a statement would be TRUE, NOT FALSE. When we complete a tree, we "read back" up the "branches" for the truth values that would make the statement true. So a final difference between tables and trees is that, while a table begins with showing all-possible combinations of truth values, a tree shows those values that would make the statement true (if it is possible for it to be true).

In this chapter, we'll begin by demonstrating the basics of tree construction using LOLA-Lite. We shall talk about using trees for statements with identity, quantifiers, and predicates in Chapter Ten.

## CONJUNCTIONS

If you're using a tree to analyze a statement, you always begin the same way. Simply *write the statement at the top of a piece of paper*. We'll use (A & B) & C as our example. Then identify the main operator, which in this case is the conjunction. Now ask yourself, **under what condition(s) would this statement be true?** For a conjunction we know there is only one condition under which it would be true: when both sides of the conjunction are true. Since we only want to write down true statements on the tree (unlike a table), beneath the statement write down the two conjuncts underneath each other; this indicates that *both conjuncts are true* when the conjunction as a whole is true. Finally, put a check beside the formula you just worked on:

A & B ✓
A
B

(A & B) & C  ✓
A & B
C

The check marks indicate that you've completed answering the question, "Under what conditions would this statement be true?" You don't have to revisit this statement.

Now look at your tree again and see whether you have any compound statements remaining. If you do, then repeat the process:

(A & B) & C  ✓
A & B  ✓
C
A
B

Now, are there any compound statements left? No. This means your tree is completed. Now read back up the branch: what you've shown is that **(A & B) & C is true if and only if A, B, and C are ALL true**.

(To double check this, complete a truth table for the statement, you should find that the statement only shows true in row 1, where A, B, and C are all true.)

Negated conjunctions are a different story. Consider the following variation on the above statement: ~(A & B) & C.

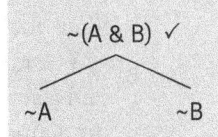

Once again, you'd begin by placing the formula at the top of the page and then breaking it down into its two conjuncts:

~(A & B) & C  ✓
~(A & B)
C

But now you have to ask yourself: under what conditions would the negation ~(A & B) be true? Since the tilde is the main connective, in order for ~(A & B) to be *true*, (A & B) would have to be *false*. So, what conditions would make (A & B) false? Again, our basic truth tables tell us that a conjunction is false if *either* conjunct is false. So, you need to have a way of representing the two different ways in which it could be false. We do this by *branching*.

Branching is when you draw one line out from the tree going to the left, and then another line out to the right. This indicates that you have two possible interpretations for what you're working on. Branching is necessary, but it obviously creates a larger, more complex tree than if you simply placed everything under the main trunk of the tree. This is further complicated if you have any untouched compound formulas above the branch. When you break those compound formulas down you have to make sure that you've placed the new material under *each* of the branches. For these reasons, when you're working on a tree, *you usually put off branching as long as you can*.

For our formula we have to branch at this point, we have no choice. What happens is that one branch covers the situation in which A is false, thereby making (A & B) false, and the other branch covers the situation in which B is false, again thereby making (A & B) false.

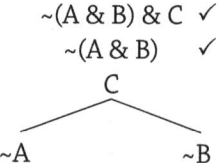

For the purposes of trees, you're done with a formula when you've broken it down into either:

a) Single statements (single capital letters)
b) Negated single statements (single capital letters with ~ in front)

When you've completed this process, you say that a tree has *terminated*.

A terminated tree will give you the possible interpretations of a statement or even a set of statements. So, in the example above you've broken down all the compound statements and you've generated two different interpretations that would make the original formula come out true. We "read up" both branches and we can see that the original statement is true when:

|  |  |  |
|---|---|---|
| C is True | **OR** | C is True |
| A is False |  | B is False |

Moreover, since *either* of these makes the statement true, you can recognize a third interpretation in which **both** A and B are false, and C is true and the formula would also come out as true.

So, when doing a truth tree if you have a conjunction, put both the conjuncts underneath it. When you have a negated conjunction, draw two branches, and put one negated conjunct on each branch. Remember to put a check by the compound statement when you've broken it down.

## Exercise 8a (Translation Review)

Remember, it's critical that you continue to work on translations in LOLA that involve quantifiers and predicates so that you don't lose your touch. We'll be re-incorporating them again very soon.

**TRANSLATION SCHEME**
RD: People

| m: Matt | c: Chad | s: Susan | t: Tim |
|---|---|---|---|
| Bx: x is a bore | Fx: x is a fireman | Hx: x is a hard worker | Dx: x is downsized |
| Gx: x is a government official | Pxy: x is promoted over y | Lx: x is a lawyer | Lxy: x is a lawyer for y |
| Sxy: x is going to sue y | | | |

1. Tim is a bore, yet he's been promoted over Susan who's not.
2. A necessary condition for being a fireman is being a hard worker.
3. Chad was downsized even though he's a hardworking fireman.
4. Sue is going to sue every government official.
5. Matt is Sue's lawyer, although he's not a hard worker.
6. There are a number of hardworking lawyers, Matt just isn't one of them.
7. Susan is going to sue Tim, even though Tim is not a government official.
8. Not all firemen are being downsized, only the hardworking ones.
9. If Matt is Susan's lawyer, then he isn't the lawyer for anyone who is a government official.
10. Tim will get promoted over Chad if and only if Chad is downsized and Susan doesn't sue every government official.

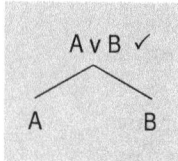

## DISJUNCTIONS

Since disjunctions are true if either of the two disjuncts is true, you'll need to branch to capture both scenarios. Consider the following conjunction that includes a disjunction as part of it: (A & B) & (D v C).

Begin the same way as you did in the previous section:

(A & B) & (D v C) ✓
A & B
D v C

Now you have a choice of which compound statement you want to work with first: the conjunction or the disjunction. As we pointed out a moment ago, you should put off branching as long as you can, but for purposes of illustration, we'll show you what it would look like if you began with the disjunction:

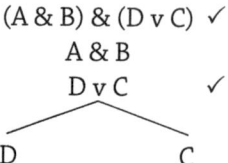

This shows the two possible interpretations, although remember that we still have another compound statement to deal with. Now we need to break down the conjunction. Since we have two different branches, *anything we do to a compound statement that occurs above a branch has to happen on both branches*. So, we do the following:

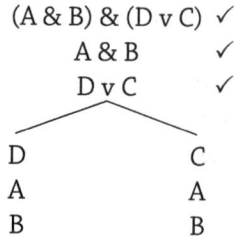

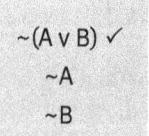

Because we've broken down all the compound statements we have completed our truth tree. We can easily see here that the original conjunction is true when D, A, and B are true OR when C, A, and B are true.

As is the case with conjunctions, constructing a tree for a *negated* disjunction is different. As always you have to begin with the main operator, which in this case is a conjunction,

$$(D \lor C) \& \sim(M \lor B) \checkmark$$
$$D \lor C$$
$$\sim(M \lor B)$$

You now know how to deal with the disjunction, so turn to ~(M v B). Ask yourself under what circumstances ~(M v B) would be true? It would be true if (M v B) is false. So, under what conditions is (M v B) *false*? We know that the *only way* that a disjunction comes out false is when *both* the disjuncts are false. So, if you were going to work with ~(M v B) first, you'd have to place a negated M and a negated B directly under the main branch, and then check off the formula.

$$(D \lor C) \& \sim(M \lor B) \checkmark$$
$$D \lor C$$
$$\sim(M \lor B) \quad \checkmark$$
$$\sim M$$
$$\sim B$$

We then move on to the remaining compound formula and break it down.

$$(D \lor C) \& \sim(M \lor B) \checkmark$$
$$D \lor C \quad \checkmark$$
$$\sim(M \lor B) \quad \checkmark$$
$$\sim M$$
$$\sim B$$
$$D \qquad C$$

We have two different possible interpretations that would make the disjunction, D v C, come out as true.

So, the rule for disjunctions is that you create two branches and put one disjunct on each branch. For negated disjunctions you negate both the disjuncts and put them both on the tree or relevant branches. The conditions under which the original statement is true is when M and B are false and D OR C is true.

## Exercise 8b

Symbolize the following statements using LOLA-Lite, construct trees for them, and then state at least one interpretation that makes the statement true.

1. Either we go to the concert or the theater, but we don't just stay home. (C: We go to the concert; T: We go to the theater; H: We stay home.)
2. The latest news is good, but we don't want to become too optimistic. (G: The latest news is good; O: We want to become optimistic.)
3. You don't go to the beach unless there are lifeguards there. (B: You go to the beach; L: There are lifeguards there.)
4. It's false that both Bob and John cheated on the exam. (B: Bob cheated on the exam; J: John cheated on the exam.)
5. It's not true that either you have to vote for a Democrat or a Republican. (D: You have to vote for a Democrat; R: You have to vote for a Republican.)
6. Each of us wants to have a successful career, personal life, good health, and not hurt other people. (C: We want to have a successful career; P: We want to have a successful personal life; H: We want to have good health; O: We want to harm other people.)
7. I don't like most television shows, unless they're those reality shows. (T: I like most television shows; R: I like reality shows.)
8. Although there can be good reasons to allow unfettered immigration, most policy analysts think the problems are too severe and the advantages too minimal. (G: There are good reasons to allow unfettered immigration; S: Most policy analysts think the problems are too severe; A: Most policy analysts think the advantages are too minimal.)
9. We either eliminate the estate tax or we are seen as tax and spend liberals, even though eliminating the estate tax is a bad idea. (E: We eliminate the estate tax; T: We're seen as tax and spend liberals; I: Eliminating the estate tax is a good idea.)
10. You can have the American dream and retain your spirituality; however it is hard work and you need to watch out that you are not tempted by all the material goods. (A: You can have the American dream; S: You can retain your spirituality; H: It is hard work; W: You need to watch out that you are not tempted by material goods.)

## BICONDITIONALS AND DOUBLE NEGATIONS

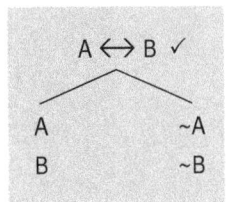

Biconditionals, like disjunctions, have two different scenarios which make them come out true. Recall that *a biconditional is true whenever both sides are true or both sides are false*. Therefore, you know that in a case where the biconditional is the main operator, you're going to have to branch right off the bat:

$$\sim(M \mathbin{\&} K) \leftrightarrow (\sim D \mathbin{v} J)$$

As always, identify the constituent parts of the compound formula before you proceed. These component parts may themselves have component parts, but you'll worry about that later. What you're looking at is the entire formula that makes up the left side of the biconditional and the entire formula that makes up the right side of the biconditional. The left side of this biconditional is ~(M & K) and the right side is (~D v J).

Therefore, one branch will have both of these statements true and the other branch will have both the statements false. This gives you the following,

$$\sim(M \mathbin{\&} K) \leftrightarrow (\sim D \mathbin{v} J) \checkmark$$

| ~(M & K) | ~~(M & K) |
| ~D v J | ~(~D v J) |

(Look carefully at this example to make sure you understand when you must retain outside parentheses and when you may omit them.)

Before we go on, take a close look at the formula on the right side, ~~(M & K), and think about what it's saying. If we were to state it in English, it would go roughly, "It is *false* that the formula, (M & K) is *not true*." You may well be saying to yourself that this is a convoluted way of saying that (M & K) is true. While this may be so, one of the rules in LOLA is that whenever you're going to negate something you do so by using a tilde, not by removing one. We'll go into this a bit more when we begin working on that branch of the tree.

Returning to the tree, we see that while we have two branches, we've broken down all the statements on the "trunk" of the tree. This means that *the branches are distinct from each other, and we treat them as separate entities*. We'll begin by working on the left branch, asking ourselves whether we can break these statements down without branching again. Unfortunately, this is not the case, so we might as well pick one as the other.

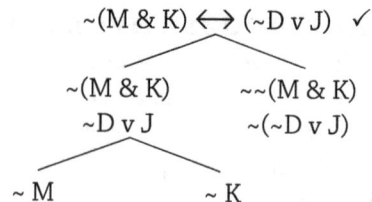

We complete the remaining formula and we're done with the left side.

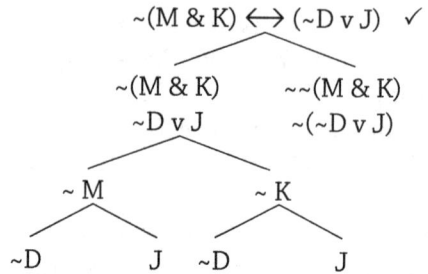

Notice that the formula ~D v J occurred above the branches that had ~M on one side and ~K on the other. So, we had to make sure that when we broke it down, all the possible scenarios were represented on all the branches lower down the tree.

This completes the left side of the tree, so we turn to the right side. Again, since we have more than one formula to work with, we look for a way to minimize branching. We have two formulas to work with ~~(M & K) and ~(~D v J). We'll begin by considering the first. Ask yourself what the main operator is in ~~(M & K). Is it the ampersand? Is it the inner tilde? The outer tilde? Since the main operator is defined as the operator that has the widest scope, and the outer tilde is the one operating over the rest of the formula, that makes it the main operator. The question then becomes, what do we do with a negated negation in a tree?

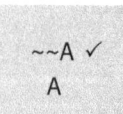

We can't simply leave it this way. Recall that when we introduced trees we said that you were done breaking down a formula when you had reduced it to either a capital letter or a negated capital letter. Fortunately we have a rule that lets us break down a double negation and it's a relatively simple one: *Whenever you have a "double negation," drop the tildes, write down the remaining statement, and then check off the original statement.*

Now we can return to our tree and check off the statement ~~(M & K) and place the new statement (M & K) underneath it.

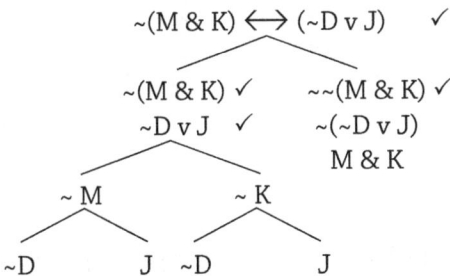

This leaves us with just two statements on the right side: ~(~D v J) and (M & K). The latter is relatively easy to treat since it's just a conjunction. The other is a negated disjunction, which means you need to negate both the disjuncts within the parentheses and place them underneath the branch. That leaves you with only one formula left to deal with, the ~~D, which you can treat according to your new rule covering double negations. The completed tree thus looks like this:

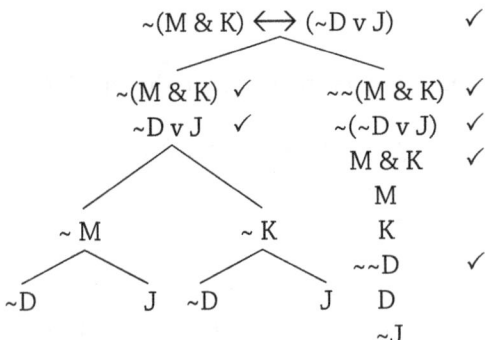

A *negated* biconditional, such as ~(A ⟷ B) *also requires branching* since there are two different ways for (A ⟷ B) to come out as false. In one scenario A is false and B is true, and in the other case, B is false and A is true. Consider the following statement:

~(A ⟷ B) v (B ⟷ ~~ E)

Having identified the disjunction as the main operator, you know you'll have to begin by branching. Having done the initial branching you then are left with a negated biconditional on the one side and a biconditional on the other side. Since they both require branching, it really doesn't matter which one you start with. We'll arbitrarily select the biconditional on the right,

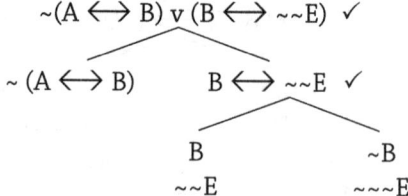

Can we do more on the right? Yes. On both sides you've got a statement that has more than one tilde on it, so apply the double-tilde rule to both sides.

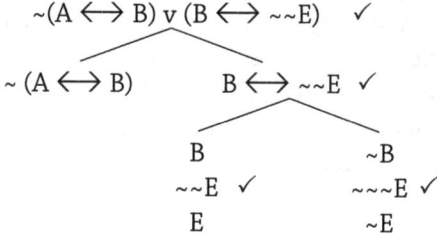

Now, you can turn to the negated biconditional on the left side.

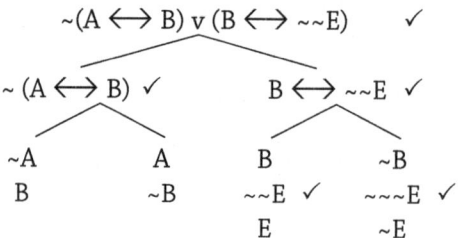

Once again, scan the branches to double check that only single statements and negated single statements remain. Seeing that this is the case, you have completed the tree.

# Exercise 8c

Symbolize the following statements using the given translation schemes and then construct truth trees for them.

1. You have fire just in case you have air, a fuel source, and a spark. (F: You have a fire; A: You have air; S: You have a fuel source; P: You have a spark.)
2. Either the dogs have been in the garbage, or someone knocked it over and isn't telling the truth. (D: The dogs have been in the garbage; K: Someone knocked over the garbage; T: Someone's telling the truth.)
3. While the Middle East is always a powder keg, there are not insignificant problems in Africa and Asia. (M: The Middle East is a powder keg; F: There are significant problems in Africa; S: There are significant problems in Asia.)
4. We support neither the destruction of the rain forest nor the removal of the multinationals from the region. (F: We support the destruction of the rain forest; R: We support the removal of the multinationals from the region.)
5. The company will support this project if and only if there is substantial interest from the public and there are no difficulties with import tariffs. (C: The company supports this project; I: There is substantial interest from the public; D: There are difficulties with import tariffs.)
6. It is not true that Sam gave a false deposition and that he is hiding something. (S: Sam gave a true deposition; H: Sam is hiding something.)
7. Either there is no new planet out there or we are receiving corrupt data from the telescope, but we can't be receiving corrupt data. (P: There is a new planet out there; C: We are receiving corrupt data from the telescope.)
8. Winning California, New York, Florida, and either Michigan or Pennsylvania is a necessary and sufficient condition for any Democrat to win the presidency. (C: A Democrat wins California; N: A Democrat wins New York; F: A Democrat wins Florida; M: A Democrat wins Michigan; P: A Democrat wins Pennsylvania; W: A Democrat wins the presidency.)
9. It's not the case that he is both incorrect and incompetent. (C: He is correct; S: He is competent.)
10. Either the oil company will succeed in obtaining the leases or the environmental lobby will stop them, but both can't happen. (O: The oil company obtains the leases; E: The environmental lobby stops them.)

Set up and complete truth trees for the following formulas:

11. ~(B & ~M)
12. K ⟷ (N v J)
13. S v (L & ~A)
14. ~~Z v P
15. ~(F ⟷ C) v ~(L & Z)
16. R & ~~C
17. {(Y & W) v Y} & (W & ~M)
18. ~~~V & (O v L)
19. U ⟷ (D & ~~ E)
20. P & ~(L v ~A)

## CONDITIONALS

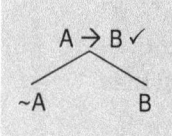

The final connective we'll examine in this chapter is the conditional. Conditionals also require branching, since *a conditional is true when either the antecedent is false or the consequent is true*. A negated conditional doesn't branch since the only way that a conditional comes out false is if the antecedent is true and the consequent is false.

Here is a statement that incorporates both,

$$(P \to M) \ \& \ \sim(Z \to R) \ \checkmark$$
$$(P \to M)$$
$$\sim(Z \to R)$$

Since we have two formulas we need to work with, the first question is whether one of these can be done without branching. Since the second one is a negated conditional, and they don't branch, we begin with that one.

$$(P \to M) \ \& \ \sim(Z \to R) \ \checkmark$$
$$(P \to M)$$
$$\sim(Z \to R) \quad \checkmark$$
$$Z$$
$$\sim R$$

Now we break down the last remaining compound statement, and we're done.

$$(P \to M) \ \& \ \sim(Z \to R) \ \checkmark$$
$$(P \to M) \quad \checkmark$$
$$\sim(Z \to R) \quad \checkmark$$
$$Z$$
$$\sim R$$

$$\sim P \qquad M$$

## OPEN VERSUS CLOSED BRANCHES AND TREES

Trees are ways of showing what interpretations would make the statement come out true. However, what would happen if we tried to construct a tree for a logical falsehood, a statement in which there is no possible interpretation that makes the statement true?

In order to understand how we treat this issue in the case of truth trees, we'll begin by making a distinction between an *open* and a *closed* branch of a tree. A *closed branch on a tree is one in which there is a contradiction on it and an open branch is one in which there are no contradictions*. It's easier to understand this by presenting some examples. Consider the following statement:

$$(M \& N) \vee [Z \& (L \& {\sim}Z)]$$

We'll set it up and go through the initial phases.

```
        (M & N) v [Z & (L & ~Z)]  ✓
              ╱          ╲
          M & N ✓      Z & (L & ~Z)
            M
            N
```

Thus far, everything is the way it has been for the earlier trees. However when we turn to the right side of the tree we're going to encounter something new. We proceed as we would normally by breaking down the conjunction and then breaking down any remaining compound statements.

This gives us

```
        (M & N) v [Z & (L & ~Z)]  ✓
              ╱          ╲
          M & N ✓      Z & (L & ~Z) ✓
            M              Z
            N            L & ~Z  ✓
                            L
                           ~Z
                            x
```

When you examine the right side of the tree you see that if you follow a branch back up the tree you'll find both a Z and a ~Z along the path. What this is saying is that if you follow this particular interpretation you will wind up trying to *simultaneously assert both that Z is true and that Z is false*. Of course, this is not acceptable according to the rules of our bivalent logic system, since *all statements in LOLA are either true or false, but can never be both at once*.

The assertion of a statement and its negation in effect asserts that that statement is *both* true and false. An interpretation which makes a statement both true and false is said to yield a *contradiction*. Any interpretation which yields a contradiction is unacceptable since it leads to a violation of a rule in **LOLA**. *In order to indicate that a particular*

> A
> ~A
> x
> (A & ~A) is an L-falsehood—
> L-falsehood kills a truth branch!!!

*interpretation has resulted in a contradiction we place an **x** under that branch and say that it is closed. If a branch terminates, but does not generate a contradiction then we say that the branch is open.*

So, the *tree of a logical falsehood should terminate with all closed branches*. Since the above tree has an open branch, that means that you could construct an interpretation for it that makes it come out true, meaning it is not a logical falsehood. Here is an example of a logical falsehood and what the resulting tree would look like:

$$(B \vee \sim B) \rightarrow \sim\sim(D \;\&\; \sim D) \quad \checkmark$$

```
      ~(B v ~B) ✓           ~~(D & ~D) ✓
         ~B                   (D & ~D) ✓
         ~~B ✓                    D
          B                      ~D
          x                       x
```

We get exactly the result that we would expect. No matter what interpretation we try, it turns into a contradiction. This means that there *is no interpretation that can make the formula come out true*, which is the definition of a logical falsehood.

There is also a way to use a tree to test for whether a formula is logically true and it entails a strategy that will be essential throughout the rest of the text. *In order to determine if some formula is a logical truth, negate the formula, and run the tree. If the original statement is a logical truth, negating it results in a logical falsehood, and the tree terminates with all closed branches.* We'll demonstrate this approach on the following formula: $(A \rightarrow B) \leftrightarrow (\sim B \rightarrow \sim A)$.

The negation of any formula is simply that formula and the addition of a tilde to it as the *main* operator. The negation of $B \rightarrow A$ is not $\sim B \rightarrow A$. It's $\sim(B \rightarrow A)$. To negate a formula, *everything about the statement must remain the same, but you put parentheses around the entire formula and place a tilde on the outside*. Once you've done that you construct the truth tree just as you always would and see if all the branches close.

Since it's a negated biconditional you'll begin by branching, then start looking for which statements you can break down without branching—in this case, the negated conditionals:

$$\sim[(A \rightarrow B) \leftrightarrow (\sim B \rightarrow \sim A)] \quad \checkmark$$

```
        A → B              ~(A → B) ✓
    ~(~B → ~A) ✓            ~B → ~A
        ~B                      A
        ~~A                     ~B
```

This leaves us with conditionals on both sides of the tree and a double negation on the left branch. When we break these down we find contradictions on all the branches.

$$\sim[(A \rightarrow B) \leftrightarrow (\sim B \rightarrow \sim A)] \checkmark$$

```
         A → B     ✓        ~(A → B)    ✓
        ~(~B → ~A) ✓        ~B → ~A     ✓
           ~B                   A
           ~~A    ✓             ~B
            A
           / \                 / \
         ~A   B             ~~B ✓  ~A
          x   x              B     x
                             x
```

Since the *negation* of our original formula results in a tree that has all closed branches, we can conclude that the original formula is a logical truth. If there had been even *one* open branch then we would know that there is at least one interpretation that could make the original formula come out false, and so we'd know it wasn't a logical truth.

We can also use trees to test whether a statement is contingent. If you construct a tree that has at least one open branch, you know the statement is true under some interpretation. But is it contingently true or L-true? We can distinguish these by using what we've just learned: Negate the original formula and run the tree. If this tree results in *all* closed branches, you know the original statement is a logical truth. If it does NOT result in *all* closed branches, the original statement is contingent.

We'll demonstrate it with the following formula:

$$(M \lor P) \mathbin{\&} (C \rightarrow M)$$

Begin by constructing a tree for the original statement:

```
     (M v P) & (C → M)  ✓
        (M v P)         ✓
        (C → M)         ✓
         /     \
        M       P
       / \    / \
     ~C   M  ~C  M
```

All branches are open. Now we need to determine if the original statement is a logical truth or contingent. So, we negate the statement and work the tree:

$$\sim[(M \vee P) \& (C \rightarrow M)] \quad \checkmark$$

```
        ~[(M v P) & (C → M)]    ✓
         /              \
    ~(M v P) ✓         ~(C → M) ✓
       ~M                  C
       ~P                 ~M
```

The negation of the original formula does NOT result in a tree with all closed branches (in fact, in this case all branches are open). Let's review what we've done and what we've learned.

1. Our *original* statement $(M \vee P) \& (C \rightarrow M)$ has at least one interpretation that comes out true.
2. This means it is *either* a contingent statement or L-true. We then ran a tree on the negation of the original statement to see which one it is.
3. The *negation* of the statement $\sim[(M \vee P) \& (C \rightarrow M)]$ has at least one interpretation that makes *it* come out true.

Because there is an interpretation that makes the negation of the original statement come out true, and at least one that makes it come out false, we know the original statement is contingent, not a logical truth. (We also knew, of course that the original formula was not logically false, because the first tree did not have all closed branches.)

If you're trying to determine whether a formula is contingent, logically true, or logically false, you can start by doing its tree or the tree of its negation. However, be clear as to what you've established when you've constructed a tree. If your tree terminates and has no open branches, then you need go no further; that statement is a logical falsehood. If your tree has at least one open branch, take the negation of the statement and see if you derive a tree that terminates with no open branches. If it does, the original statement is L-true; if it doesn't, the original statement is contingent.

## Exercise 8d

Symbolize the following statements using the given translation schemes and then construct truth trees for them.

1. If the boat comes in early, we'll be able to unload the fruit quickly, though we won't be able to do anything about the rest of the cargo. (B: The boat comes in early; F: We unload the fruit quickly; C: We can do something with rest of the cargo.)
2. We will move to Syracuse if and only if I can get a raise and an upgrade in my ranking. (S: We move to Syracuse; R: I get a raise; U: I get an upgrade in my ranking.)
3. Both Aristotle and Plato stress the importance of form in their ontology, although they disagree with each other about what it means. (A: Aristotle stresses the importance of form in his ontology; P: Plato stresses the importance of form in his ontology; I: Plato and Aristotle agree on what form means.)
4. Either you find the key to the map, or we stop and ask someone for directions. (K: You find the key to the map; S: We stop; D: We ask someone for directions.)
5. Assuming it doesn't rain, we can work on the project and make some progress. (R: It rains; P: We can work on the project; M: We make some progress.)
6. A necessary condition for this being a mental event is that it be non-spatial and non-temporal. (E: This is a mental event; S: This is spatial; T: This is temporal.)
7. It's not true that if I can't do this exercise I'm out of shape. (D: I can do this exercise; I: I'm in shape.)
8. The international press had a field day with this story, however we're either misreading the polls or it turns out that there has been no damage on the domestic front. (P: The international press had a field day with this story; C: We're correctly reading the polls; H: There has been damage on the domestic front.)
9. Thomas Aquinas' five proofs for the existence of God work just in case you accept a particular understanding of the nature of causation, which I don't. (T: Thomas Aquinas' five proofs for the existence of God work; A: You accept a particular understanding of the nature of causation; I: I accept the particular understanding of the nature of causation.)
10. There must be a practical way to solve the problem of rampant poverty or we have to accept that poverty is endemic and we are helpless to do anything about it. (P: There is a practical way to solve the problem of rampant poverty; E: We accept that poverty is endemic; H: We are helpless to do anything about poverty.)

For the following, construct truth trees to determine if they're contingent, L-true or L-false.

11. ~(A & B) & (A & B)
12. (P → M) → (~P v M)
13. (M & N) → ~(N & M)
14. ~(Z & L) ↔ (~Z v ~L)
15. [(D & C) → M] → ~[D → (C & M)]
16. (Z & K) v (~J & ~O)
17. (R ↔ Q) → [(R → Q) & (Q → R)]
18. ~(Y v W) → (~Y & ~W)
19. [M & (M → L)] & ~L
20. (~Y & ~W) → ~(Y v W)
21. ~[(~F v H) → (F → H)]
22. ~[K & (K v O)]
23. [(U v G) & ~G] → U
24. ~[Z & (Z v J)]
25. ~[{I v (T & V)} ↔ {(I v T) & (I v V)}]
26. [(R & T) v (K → ~R)] & ~(R & T)
27. ~~[Y v (P → ~Y)]
28. [(A & C) → ~C] ↔ (A v C)
29. ~(N & M) → (M & N)
30. [(~P v M) & (L v M)] → (M v ~P)

## Exercise 8e (Translation Review)

Using the given translation scheme, give complete translations in LOLA, NOT LOLA-lite.

**TRANSLATION SCHEME**
RD: Everything

| a: John Adams | j: Thomas Jefferson | b: Aaron Burr | h: Alexander Hamilton |
|---|---|---|---|
| Px: x is a person | Ox: x was a president | Sx: x was Secretary of State | Ex: x was an excellent shot |
| Kxy: x killed y | Dxy: x fought a duel with y | Bxy: x beat y in the election | Ixy: x insulted y |

1. Fighting a duel is a symmetrical relationship.
2. Aaron Burr insulted many people, but he fought a duel with Alexander Hamilton who insulted him.
3. Jefferson beat Adams in the election, but they did not insult each other.
4. Both Jefferson and Adams were president, but Burr and Hamilton weren't.
5. It's not true that Hamilton was an excellent shot and, not surprisingly, Burr killed him.
6. Everyone insults someone, but not everyone fights a duel with someone.
7. No one beat Jefferson in the election, though Adams did beat Burr.
8. Adams did not kill Jefferson or anyone else for that matter.
9. Jefferson was Secretary of State and President, but neither Burr, nor Adams, nor Hamilton was ever Secretary of State.
10. Jefferson and Adams were excellent shots, but in fact they did not duel with anybody.

# Chapter Nine: Relationships between Statements

*What's Up?*
Consistent
Contrary
Contradictory
Implication
Equivalence

Thus far we have focused on single statements. From here on we will extend the analysis to include examination of the *relationships* that can obtain between and among statements. In this chapter we introduce you to five logical concepts governing the relationship between two statements: consistency, contrariety, contradictoriness, implication, and equivalence. With each one we first introduce the concept via truth tables, and then we'll show it again with truth trees. We hope that by doing so it will deepen your understanding of each concept, as well as strengthen your skills with both techniques.

## CONSISTENCY

*Two statements are consistent if there is at least one interpretation that makes them both true.* As you'll recall from previous chapters, an interpretation is an assignment of trues and falses to the atomic statements. To say that two statements are consistent means you've got at least one type of situation—that is, one way true and false are distributed among their atomic sentences—in which they both are true. There may also be situations where one of them is true and the other false or where both come out false, but that doesn't matter. All it takes is a single scenario that makes both statements true for them to be consistent. We'll first show this to you with the following statements and a truth table.

> **Example Set 9a**
> 1. We went to the store and had the car fixed. (S: We went to the store; C: We had the car fixed.)
>    Translation: S & C
> 2. We went to the movies and had the car fixed. (M: We went to the movies; C: We had the car fixed.)
>    Translation: M & C

**Consistent Statements**
Possible for both statements to be true at the same time.

To show that these two statements are consistent we construct a truth table, although with a slight difference from the previous ones. This time, since we're exploring the relationship between both statements we have to count the atomic statements from *both* statements when calculating the number of rows and columns. We have C, M, and S, which means we'll need to construct a table with one row on top and eight rows underneath it. Furthermore, in addition to including columns for all the atomic statements, we'll need columns for *both* compound statements.

For the example above, this means we'll need five columns. Here's what the table would look like.

| C | M | S | S & C | M & C |
|---|---|---|-------|-------|
| T | T | T | T | T |
| T | T | F | F | T |
| T | F | T | T | F |
| T | F | F | F | F |
| F | T | T | F | F |
| F | T | F | F | F |
| F | F | T | F | F |
| F | F | F | F | F |

*Once you have completed the table you look at the final columns and see if you have a row in which both WFFs come out as true.* In this case, the first row has T under the main operator in both WFFs. What this means is that we can construct an interpretation in which both WFFs are true. What is more, we can simply read off what this interpretation is. If you read the first row from left to right you see that whenever

C is True
M is True
S is True

then both the statements come out as true. This *tells us that the two formulas are consistent*. If you do any other assignment of Ts and Fs you wind up with a situation in which at least one of the WFFs is false.

We can also check for consistency between these two statements with truth trees. Since we want to know if it is possible for both statements to be true under one interpretation, we *put both statements, one after the other, at the top of the tree*. Then we finish the tree as usual:

```
        S & C  ✓
        M & C  ✓
          S
          C
          M
          C
```

Since the tree terminates with at least one open branch, it means that there is an interpretation that makes the WFFs both come out true. And as in the truth table, we can see that both statements are true when S, C, and M are all true.

## CONTRARIES AND CONTRADICTORIES

There are three terms that sound a bit alike and are easy to confuse: contradiction, contrary, and contradictory. You have already worked with the concept of contradiction in the previous chapter. A branch that has both a statement and the negation of that statement is understood as an attempt to be asserting both statements at the same time. We call this a contradiction and we close off the branch it occurs on as an untenable interpretation.

> ***Contradictory Statements***
> Statements that always have opposite truth values.

*Contradictories*, on the other hand, are two statements that *always have opposite truth-values* under any interpretation. When statements are in this relationship, we say that the statements are *contradictory*. We'll demonstrate this with the following statements, first using a truth table, and then a truth tree.

---

**Example Set 9b**
1. If Paul goes to medical school, then he'll make a lot of money. (P: Paul goes to medical school; M: Paul makes a lot of money.)
   Translation: (P → M)
2. It's just not true that Paul makes a lot of money unless he doesn't go to medical school. (P: Paul goes to medical school; M: Paul makes a lot of money.)
   Translation: ~(M v ~P)

---

Construct a truth table in the same way that we did earlier when we were introducing the concept of consistency. Because we only have two atomic statements this time, the truth table will only have four rows under the top row and only four columns. For reading ease, we'll just fill in the final columns under the main connectives (you would, of course, want to fill in all the columns were you doing the table).

| P | M | P → M | ~(M v ~P) |
|---|---|-------|-----------|
| T | T | T     | F         |
| T | F | F     | T         |
| F | T | T     | F         |
| F | F | T     | F         |

This table shows no assignment of Ts and Fs where the two statements come out true at the same time—they're **in**consistent. And they always have opposite truth values—they're contradictories.

We can also determine that two statements are contradictory with truth trees. It helps to think through what we're trying to show. First, contradictories always have opposite truth values, that is, they can't both be true and they can't both be false. If they can't both be true, then we know they're *not consistent*. To show this, we need to do a consistency tree, as we did above. Then we'd need to check if they both can be false. How? We construct a tree where both statements are negated at the top of the tree.

Since we have to do both steps (that is, check if consistent, then check if contradictory) we can set up one tree with two branches. At the top of the tree we construct a biconditional with one of the statements to the left of the double arrow, the other statement to the right of the double arrow. For our example statements, the top of the tree looks like this

$$(P \rightarrow M) \longleftrightarrow \sim(M \text{ v } \sim P)$$

Recall that a biconditional is only true when the two components of it have the same truth-value. Hence, a biconditional is true if and only if both sides are true or both sides are false. So the tree branches: left branch with both statements true; right branch with both statements false. And it looks like this:

$$(P \rightarrow M) \longleftrightarrow \sim(M \text{ v } \sim P) \checkmark$$

(P → M)                    ~~(M v ~P)
~(M v ~P)                  ~(P → M)

The left branch is the "consistency branch," the right branch is the "inconsistency branch." If the two statements are contradictories, the consistency branch should terminate with all closed branches (they can't both be true) AND the inconsistency branch should also terminate with all closed branches (they can't both be false). Let's finish the tree and see:

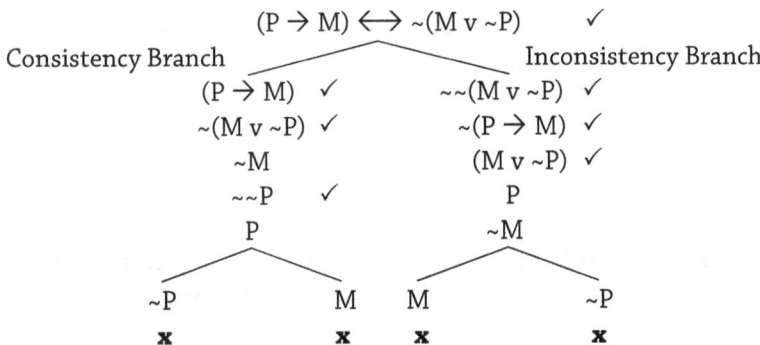

All the branches indeed close. These statements are **CONTRADICTORY**.

Two statements can also be *contraries*. Contrary statements are *inconsistent* statements, that is, they both can't be true. But they both can be false. In other words, for any two statements, A and B, that are *contraries*, you have only one of three possibilities:

A is true, but B is false **OR**
B is true, but A is false **OR**
Both A and B are false.

> **Contrary Statements**
> Impossible for both statements to be true at the same time, but both can be false at the same time.

We'll use the following two statements to demonstrate how you would establish this, first with tables and then with trees.

---

**Example Set 9c**
1. I figure that if you wanted pineapple or cantaloupe, you didn't want pineapple. (P: You want pineapple; C: You want cantaloupe.)
   Translation: (P v C) → ~P
2. You want both pineapple and cantaloupe. (P: You want Pineapple; C: You want cantaloupe.)
   Translation: (P & C)

---

To demonstrate contraries with a truth table all you need to do is to take the two formulas and place them in a truth table the same way you have with the previous concepts. Again, for ease of reading, we'll just put down the truth values of the final columns.

| C | P | (P v C) → ~P | P & C |
|---|---|---|---|
| T | T | F | T |
| T | F | T | F |
| F | T | F | F |
| F | F | T | F |

When we read the truth table, we see that there are interpretations in which a formula comes out true, but in each case, the other formula comes out false on that same row. We also have a row in which they both come out false. However, we have no row in which they both come out true.

To express this through truth trees, let's think a moment about what we've just shown. A contrary relationship is where the statements can't both be true, in other words, they are *not* consistent, but they could both be false. So, we can show the contrary relationship by setting up our "biconditional tree," just as we did for contradictories. Complete the consistency (left) branch; if all branches close they are *in*consistent. Proceed to the inconsistency (right) branch. Unlike contradictories, contrary statements *can* both be false at the same time, so we should find at least one open branch on the right-hand branch.

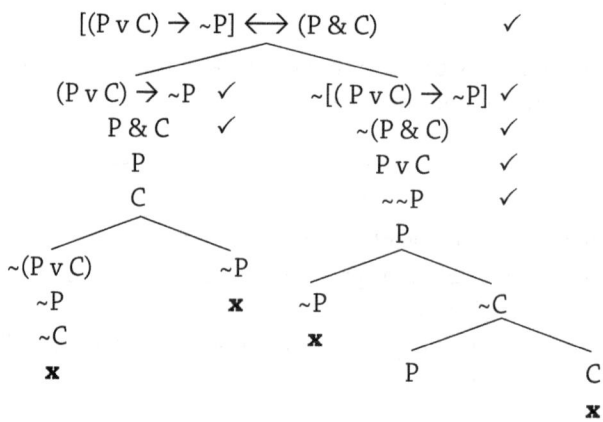

Consistency branch has all closed branches: statements are inconsistent.
Inconsistency branch has an open branch: statements are CONTRARY.

# Exercise 9a

Run both a truth tree and a truth table for the following pairs of propositions and explain why the propositions are consistent, contraries, or contradictories.

1. ~(A & B), M → (B v ~M)
2. ~(P v C) v ~P, ~(~P v ~C)
3. ~(M v N) v ~M, ~(~M v ~N)
4. ~[(A & B) → C)], A → (B → C)
5. L & (Q v D), ~{(L & Q) v (L & D)}
6. Z → K, ~K v (Z & L)
7. ~[(W v L) & L], L ↔ W
8. A & B, B v C
9. ~[(P v C) & P], P & C
10. [(N v J) & ~N], J v ~N
11. D ↔ R, [~(D → R) & (R → D)]
12. (F → G) & ~G, (G → F) & ~F

# Exercise 9b (Translation Review)

In the next chapter, we'll be re-introducing quantifiers, predicates, names, and so on, so make sure you keep doing these translation reviews along the way. Use full LOLA!

**TRANSLATION SCHEME**
RD: Everything

| q: the McLeod quartz | d: the Hope diamond | r: the Delong Star ruby | s: the Star of India |
| --- | --- | --- | --- |
| | | | f: the Jackson flintrock |
| Kx: x is well known | Px: x is a person | Fxy: x is more famous than y | Wx: x is worthless |
| Axy: x can afford y | Wxy: x is worth more than y | Hxy: x has heard of y | Cxy: x causes y to go mad |

1. The Hope diamond is worth more than the McLeod quartz, but no one can afford the Hope diamond.
2. The McLeod quartz is worthless, but it's still worth more than the Jackson flintrock.
3. The Star of India is famous only if everyone has heard of it.
4. If no one has heard of the Jackson flintrock, then it's worthless.
5. The Delong Star ruby caused everyone to go mad.
6. Being famous is a transitive property.
7. The McLeod quartz makes no one go mad.
8. Either someone has heard of the Delong Star ruby or no one has, but it can't be both ways.
9. The Delong Star ruby and the Star of India are both well known, but neither is as famous as the Hope diamond.
10. Nothing is worth more than the Star of India unless it causes everyone to go mad.

> **Implication**
> A 1st statement implies a 2nd statement if it's impossible for the 2nd statement to be false when the 1st is true.

## IMPLICATION

The next relationship between statements we'll discuss is called implication. To say that one statement implies another statement means that *it's **not** possible for the second statement to be false when the first statement is true*. Consider the following two statements.

---

**Example Set 9d**
1. Jennifer wants to go to New York and see a show. (N: Jennifer wants to go to New York; S: Jennifer wants to see a show.)
Translation: N & S
2. Jennifer wants to see a show. (S: Jennifer wants to see a show.)
Translation: S

---

It's essential with implication that you're clear about which statement you're saying is doing the implying and which statement is being implied. We can represent this relationship symbolically with the symbol ∴ . Whatever is on the left of this symbol is doing the implying and whatever is on the right side of it is what's being implied. So, if we want to say that sentence 1 implies sentence 2 we would write it like this:

$$N \& S \therefore S$$

Be clear. This *does not say that S implies N & S*. That may or may not be the case, but all this sentence commits us to is saying that N & S implies S. Again, that means that it's impossible for N & S to be true and for S to be false. It's easy enough to show this with a truth table, although there is one new feature to the truth table that has to be added to show that we talking about the implication relationship.

Set everything up as you would normally in a truth table, but when you're drawing the vertical line to separate N & S from S, make it a double line. Your truth table should look like this,

| N | S | N & S | S |
|---|---|-------|---|
| T | T | T     | T |
| T | F | F     | F |
| F | T | F     | T |
| F | F | F     | F |

By reading from left to right you can see that there is no row in which the statement, N & S, is true where it isn't also true under the statement S.

Of course, you can also use the truth table to ascertain whether one statement ***doesn't*** imply the other. Simply put S to the left of the double line and N & S to the right of the double line and read the resulting truth table. When you do, you find that in row 3, S reads as true, but N & S reads as false. These two tables establish that N & S implies S, but that S *does not* imply N & S.

| N | S | S | N & S |
|---|---|---|---|
| T | T | T | T |
| T | F | F | F |
| F | T | T | F |
| F | F | F | F |

Another example:

Statement 1: ~(~A v A)
Statement 2: ~A → B

Does Statement 1 ∴ Statement 2? Below is the complete truth table for these two statements. Take a few minutes to study the table (before reading the analysis below) and then decide if the 1st ∴ 2nd.

| A | B | ~(~A v A) | ~A → B |
|---|---|---|---|
| T | T | **F** F T | F **T** |
| T | F | **F** F T | F **T** |
| F | T | **F** T T | T **T** |
| F | F | **F** T T | T **F** |

The final column of Statement 1 shows all Fs in the final column—we can see that Statement 1 is logically false. The final column of Statement 2 has both Ts and Fs. But does Statement 1 imply Statement 2? Remember the definition of implication: it's impossible for the 2nd statement to be false when the 1st is true. But what if the first statement is never true, as in our example? Another way to state the implication relation is *there will never be the case where the first is true and the second false*. And this is just what our table shows: there's never a case where the first is true and the second false. So, for our example:

**Statement 1 ∴ Statement 2**

Implication, the way we're thinking about it, is importantly different from what you might think of when it's said that one statement implies another, or that the second logically follows from the first. According to *our* notion of implication, A & ~A ∴ B—so Arthur is tall and Arthur is not tall implies Brazil is in South America. It's the same sort of case as the one we've just looked at: the first statement is logically false. So a logically false statement implies all statements!

Truth trees can also be used to determine implication. Since what you're trying to learn is whether it's impossible to have an interpretation in which the one statement can be true and the other statement is false, you set up a tree that looks just like that. You place *the 1st statement on the top and then the negation of the 2nd statement underneath it*. You then run a tree and see whether it terminates with any open branches or if all the branches close. *If all the branches close, then you've shown that it's not possible to construct an interpretation in which the first statement is true and the second statement is false*, which means that the first one implies the second one. If the tree terminates with an open branch, then you've shown it *is* possible to construct such an interpretation and, hence, the first statement does not imply the second.

---

**Example Set 9e**
1. John is a great guy. (J: John is a great guy.)
   Translation: J
2. John is a great guy or John is obnoxious. (J: John is a great guy; O: John is obnoxious.)
   Translation: J v O

---

To determine whether J ∴ J v O, set up the tree, making sure to put the *negation* of the second statement underneath the first statement.

$$
\begin{array}{c}
J \quad \checkmark \\
\sim(J \lor O) \quad \checkmark \\
\sim J \\
\sim O \\
x
\end{array}
$$

Since there is a contradiction on the branch, you put an x down indicating it's closed. This means that the tree has terminated with no open branches, thereby establishing that there's *no way for the first statement to be true and the second statement to be false*. Hence, the first statement *does* imply the second.

Just as with tables you can also use trees to show that one statement *doesn't* imply another. We'll simply reverse the statements so that J v O is the first one, without negation, and then ~J , below that, is the negation of the other statement. When we set the tree up we have the following:

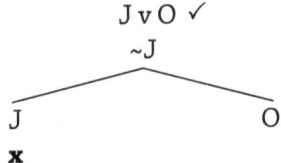

x

In this case, while the left branch closes, the right side is still open. This shows it *is* possible for the first to be true and the second false. Thus, we've established that while J implies J v O, J v O *doesn't* imply J.

## Exercise 9c

Use both trees and tables to demonstrate whether the first formula implies the second formula. If it doesn't, then use either tables or trees to go on to show whether the statements are consistent, contraries, or contradictories.

1. M & (B v C), C → ~B
2. [A ↔ (D & ~~A)], ~~~A
3. Z → Z, W v ~W
4. (L v U), ~(~L & ~U)
5. (R & S) → (K & R), R → K
6. O, O v (N v ~N)
7. Z v (M & Y), (Z & M) v (Z & Y)
8. P → ~Q, ~(Q → ~P)
9. J ↔ D, ~J & D
10. F v (W v P), (F v W) v P

## EQUIVALENCE

The last concept we'll introduce in this chapter is equivalence. To say that *two statements are equivalent means that they always have the same truth value.* You can think of equivalence as the opposite of contradiction (where they never have the same truth value), and the way you establish equivalence is quite similar to how you establish contradiction. For tables, set up the two statements as usual and then read the table to see whether there is any row in which there are different truth-values under the main connectives of the respective statements.

> **Equivalent Statements**
> Two statements are equivalent if they each imply the other.

### Example Set 9f
1. The union will go on strike only if management pushes for increases in co-pay rates. (U: The union goes on strike; M: The management pushes for increases in co-pay rates.)
   Translation: U → M
2. If the management won't push for increases in co-pay rates, then the union won't go on strike. (U: The union goes on strike; M: the management pushes for increases in co-pay rates.)
   Translation: ~M → ~U

Set up the table as you would normally (again for readability we just give final-column values).

| M | U | U → M | ~M → ~U |
|---|---|---|---|
| T | T | T | T |
| T | F | T | T |
| F | T | F | F |
| F | F | T | T |

As you can easily see, in each row, the truth value for one statement is the same as the truth value for the other. So, these two statements are equivalent.

Truth trees can also be used to establish equivalence, but it's a bit more complicated. Another way to think about equivalence is that you're saying that the *two formulas mutually imply each other*. So, if you show that the first formula implies the second, and then show that the second formula implies the first, you'll have demonstrated that the two formulas are equivalent. This means that, unlike tables where you can demonstrate the equivalence of the formulas in a single table, the tree method requires the construction of two trees.

**Example Set 9g**
1. If Bob has lost all his memories, then it's highly likely that his cerebral cortex is damaged. (C: The cerebral cortex is damaged; L: Bob has lost all his memories.)
Translation: L → C
2. Either Bob hasn't lost all his memories or his cerebral cortex is damaged. (L: Bob has lost all his memories; C: The cerebral cortex is damaged.)
Translation: ~L v C

Let's begin with determining if L → C ∴ ~L v C. We set this up just the way that we did in the case of implication, with the first statement on top and the negated second statement underneath it, and then see if the tree terminates with all closed branches.

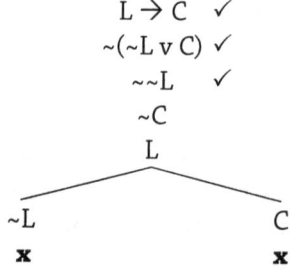

Since the tree has only closed branches, then you've shown that there can be no interpretation in which

L → C is true, but ~L v C is false.
So, **L → C ∴ ~L v C**.

To show equivalence, you now need to do another tree with ~L v C on top and the negation of L → C underneath.

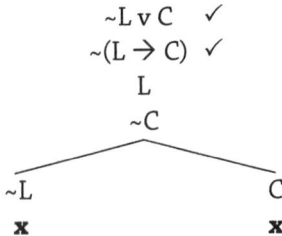

Since this tree also terminates with all closed branches, you've shown that *the two statements mutually imply each other, which means that they are equivalent.*

## Exercise 9d

Use both trees and tables to establish whether the first statement implies the second. If it does, then determine whether the statements are equivalent. If the first statement doesn't imply the second, use either trees or tables to demonstrate whether the statements are consistent, contraries, or contradictories.

1. P & (P → Q), Q
2. (M v Z) v J, J v (M v Z)
3. (V & L) → ~(L v K), L v ~ L
4. F ↔ M, (F & M) v (~F & ~M)
5. (A → B) & (B → C), ~(A → C)
6. ~(~P → C) v ~P, ~(P → ~C)
7. Q v R, ~(~R → Q)
8. N & W, W & (N v ~W)
9. ~~~~J, ~~~J
10. U ↔ N, (N v U) & (~N v ~U)
11. A → (B → C), ~[(A & B) → C]
12. (D v E) & (D → E), E
13. ~(G & P), ~G v ~P
14. J ↔ D, ~(J v ~D)
15. (W & Y) → Z, W → (Y → Z)

# Chapter Ten: Reintroducing Names, Predicates, Quantifiers, and Identity

> **What's Up?**
> Logical Properties and Relations with full LOLA

In the past three chapters we used LOLA-Lite for our level of analysis of propositions, and so we've ignored much of the complexity of our language. This is problematic since the concepts we introduced in Chapters Eight and Nine (logical truth, logical falsehood, contingency, contradiction, contradictoriness, implication, and equivalence) are also applicable to the material we presented in the first six chapters. Simplifying in this way, however, did allow us to focus on the truth-functional connectives, and to define them via truth tables, but truth tables don't work when it comes to quantified or identity statements. Fortunately, truth trees do not suffer from this drawback. In this chapter we will explain how to use truth trees with the more complex statements of the kind we worked with in the first six chapters.

## CONTINGENCY, L-TRUTH, L-FALSEHOOD

We'll begin with the logical properties that apply to a single statement: contingency, logical truth, and logical falsehood. Consider the following example set.

---

**Example Set 10a**
TRANSLATION SCHEME
RD: Numbers

| t: two | Px: x is a prime | Ex: x is even | Sx: x is small |

1. If two is a prime number then it's an even prime number.
2. Two is a small prime number, but not a prime number that is small.
3. Two is either an even number or it's not.

We begin by translating into LOLA, starting with sentence 1.

  Translation: Pt → (Et & Pt)

We then identify the main connective, which in this case is the arrow, and begin the tree. Later in the chapter we will explain how to do a tree on a formula in which the main operator is a quantifier.

$$
\begin{array}{c}
\text{Pt} \to (\text{Et \& Pt}) \checkmark \\
\diagup \qquad \diagdown \\
\sim\!\text{Pt} \qquad \text{Et \& Pt} \checkmark \\
\text{Et} \\
\text{Pt}
\end{array}
$$

Since we will no longer have single capital letters standing alone, we need a new way to determine if a branch has closed. Essentially, we'll be using the same rule in that a tree branch closes when you encounter a statement and its negation on that branch. While for the past three chapters we've said that would be a single letter and its negation, you now have a contradiction whenever you have a predicate coupled with a name, and the negation of that coupling. You don't have that in the above tree, which means that it's possible to construct an interpretation of the statement in which it comes out true. Just as before, you could follow the branches and find out exactly what interpretation would make it so. In this example you actually have several possible interpretations. Here are three possible interpretations that you can see from the tree.

> Interpretation #1: If you move along the left branch you see that if Pt is false, then, regardless of the values of the other atomic statements, the statement as a whole is true.

> Interpretation #2: If you move along the right branch you see that if Et and Pt are both true, then the statement as a whole is true.

> Interpretation #3: Since both branches are open it leaves open the possibility of using both interpretations. So, if Pt is false, and Et and Pt are both true then the statement as a whole is true.

> At the least this means that the statement is not logically false. In order to know if it is contingent or if it is a logical truth, we use the same method as before. We negate the statement and run another truth tree. If it closes, then that means the

original statement is a logical truth. If the second tree has at least one open branch, then the statement is contingent, because there is at least one interpretation in which the negated statement comes out true (so, there's at least one interpretation where the original statement can be false).

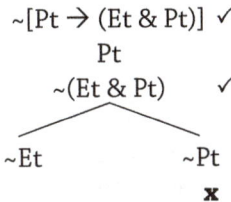

Closed Branch
Lrj
~Lrj
x

~[Pt → (Et & Pt)] ✓
Pt
~(Et & Pt)   ✓

~Et          ~Pt
              x

While one of the branches does close, there is still one that remains open. This means there is an interpretation in which this statement comes out true, hence, the original statement is contingent, and not a logical truth.

When we symbolize sentence 2 we get (St & Pt) & ~(Pt & St). When we run a tree on it we obtain the following result:

(St & Pt) & ~(Pt & St) ✓
St & Pt          ✓
~(Pt & St)
St
Pt

~Pt          ~St
x            x

Because all branches close, we know that the statement is a logical falsehood. There is no interpretation that makes it come out true.

Sentence 3 symbolized is Et v ~Et and the resulting tree shows all open branches.

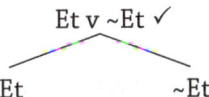

Et v ~Et ✓

Et          ~Et

By now, for this simple tree, you will probably say this is obviously L-true, because you can *see* that the statement Et v ~Et is true when either Et is true or Et is false—in other words, under any possible interpretation. Not all L-truths are so easy to see. But let's run a second tree on its negation, to demonstrate clearly that the statement is indeed L-true.

~(Et v ~Et) ✓
~Et
~~Et ✓
**x**

As expected, the tree terminates with no open branches so we can see that the original statement is a logical truth.

## Exercise 10a

Symbolize the following sentences and then use trees to ascertain whether they are logically true, logically false, or contingent.

**TRANSLATION SCHEME**
RD: Everything

| a: Aristotle | p: Plato | s: Socrates | Ax: x is an Aristotelian |
|---|---|---|---|
| Px: x is a Platonist | Sx: x is a Sophist | Rxy: x was more religious than y | Wxyz: x was more wrong than y about z |

1. It's false that Plato was more religious than Socrates.
2. A sufficient condition for Plato being a Platonist is that he not be either a Sophist or an Aristotelian.
3. Aristotle was not a Sophist only if he was a Platonist.
4. If Socrates is both a Sophist and not a Sophist, then he's also a Platonist.
5. Plato was not a Platonist, though he was more religious than Aristotle.
6. If it's false that Socrates was both a Sophist and not a Sophist, then we should also say he was either a Platonist or a non-Platonist.
7. Plato was an Aristotelian if and only if he was not a Sophist.
8. It's a necessary condition for Aristotle being an Aristotelian that he not be a Platonist.
9. Plato was more wrong about Socrates than Aristotle.
10. Socrates, Plato, and Aristotle were not Sophists.

Determine whether the following propositions are logical truths, logical falsehoods, or contingent statements. Then using the same translation scheme as above, translate them into English sentences.

11. ~Sa
12. ~Ps → (Pp v Ps)
13. Rsp & (As & ~Ps)
14. (Ap → ~Aa) & (Aa & Ap)
15. Wpas → ~Waps
16. (Rap v ~Rap) → (Ap & ~Ap)
17. (Sp & ~Sp) → (Ps v ~Ps)
18. Wpsa & Rpa
19. Ss v (Rsa & Rsp)
20. Ps & (Ps v Pa)

## QUANTIFIERS AND CONSISTENCY, CONTRADICTORIES, AND CONTRARIES

As you can see from the first section, as long as the main operator in a statement is a tilde, arrow, ampersand, wedge, or double arrow, constructing a tree is no different from the process we first introduced in Chapter Seven. However, quantifiers require us to introduce four new rules to deal with the following kinds of statements:

1. Universal Statements
2. Negated Universal Statements
3. Existential Statements
4. Negated Existential Statements

We'll use consistent, contradictory, and contrary statements to introduce these new rules. Recall that to say that two statements are consistent means that there is at last one interpretation under which they both come out true. Contradictories are statements that always have opposite truth-values. Contraries are statements that both can't be true, but both can be false. Consider the following example set.

### Example Set 10b
**TRANSLATION SCHEME**
RD: People

| b: Bob | r: Ray | Hx: x is happy | Fx: x is free |
|---|---|---|---|

1. Someone is happy. Bob is not happy.
2. Bob and Ray are happy. All free people are happy.
3. If someone is free, then someone is happy. No one is happy.
4. Someone is not free and not happy. Everyone is happy.

The first pair of statements would be symbolized as follows:

$$(\exists x)Hx, \sim Hb$$

To determine whether they are consistent you set up a tree with one formula on the top and one below it, and see if the resulting tree terminates with at least one open branch. Before you can do this, though, you need to have a rule for dealing with existential statements.

> (∃x)(Px & Lxb) ✓
> Pa & Lab
> 'a' new to branch

An existential statement says that *something* exists of a particular sort. However, we can't say what exactly that thing is. If we say that "Someone is happy" then all we're really allowed to say is that somewhere out there is some person who's happy. It's fairly common, however, for us in English to expand on that sentence in the following way, "Someone, let's just call him Allen, is happy." In such an expression, we're picking *a name at random* and making it clear that this is what we're doing. We're not actually asserting that the individual being referenced in the existential statement really is the guy named Allen. Hence, we shouldn't select the name of someone or something who we're already talking about. If we did that, then the implication would be that we're not really picking the name at random, but that we're making an assertion about this particular individual.

This process of dropping a quantifier and replacing all the variables bound to that variable is called *creating an instance*. The rules governing universal statements and existential statements differ, so we'll cover universal statements in a moment. One restriction that is the same for both existential and universal quantifiers, however, is that you *can only create an instance when the main operator is a quantifier*. You can't create an instance for a statement whose main operator is an arrow or a conjunction or an identity statement, for example. This means you may have to wait until later in a tree to create an instance.

With this in mind we can construct a rule for breaking down existential quantifiers in a truth tree. When you come to an existential quantifier in a truth tree, you drop the quantifier and replace all the variables that are bound to that quantifier with an individual constant (name) of your choosing. *The only restriction is that you can't choose an individual constant that already occurs anywhere along that branch*. So, in our example above you'd run the tree as follows,

$$(\exists x)Hx \checkmark$$
$$\sim Hb$$
$$Ha$$

"Ha" is an instance of the existential statement "$(\exists x)Hx$." The only name you *couldn't* have chosen was "b" since that's the only one that occurs on the branch already. Since the tree terminates with at least one open branch it means that our original two statements are consistent.

The second set of statements would be symbolized as follows:

**Hb & Hr, (x)(Fx → Hx)**

and the corresponding tree would be set up as follows:

CHAPTER TEN: REINTRODUCING NAMES, PREDICATES, QUANTIFIERS, AND IDENTITY

>Hb & Hr ✓
>(x)(Fx → Hx)
>Hb
>Hr

As always you want to make choices so that you don't increase the number of branches if at all possible. Hence, it's a good idea to do the conjunction first. This leaves you with a universal statement and a bit of a puzzle. Since a universal statement makes a claim about *everything*, then how will we be able to break it down into individual names? We can't put down the names of everything under the statement, or we'd never finish.

To understand better what you're to do with a universal statement, we need to think a little bit about what we're trying to do when it comes to truth trees. The strategy behind a truth tree is *to get the tree to terminate with all closed branches if possible*. Of course you have to follow the rules strictly, but you're always on the lookout for whatever "legal" moves you can make that will close off a branch. With this in mind, we can construct the rule for breaking down universal statements.

>~Wa
>(x)(Ax → Wx) *
>Aa → Wa

*When you have a universal statement put a star instead of a check by the universal statement.* This indicates that you're going to create an instance, but that this universal statement is never actually finished. You may need to go back to it at a later point in the tree and derive another instance from it. To determine what individual constant you'll choose, look over the branches and see what ones have already been used. Then create an instance or instances on that branch *using those individual constants*. You may well have to go back several times to create the required statements.

Some of you may already see that this new rule could lead to a substantial amount of work, since every time a new individual constant crops up on a branch, you'll have to go back to the universal statement above it, and create a new instance which incorporates this new individual constant. At your instructor's discretion, we will permit a certain amount of latitude in implementing the universal statement rule. As you become more familiar with trees, you will start to see that creating a particular instance of some universal statement couldn't possibly close the branch. At the instructor's discretion, you will be allowed to skip putting a particular instance of the universal on a branch if there isn't any way that creating that instance on a particular branch will actually close it. Remember, though, that depending on what happens on the various branches, you may need to go back to that universal statement and place that instance of it on a branch that now could be closed if you did so. Let's return now to the tree we began a moment ago:

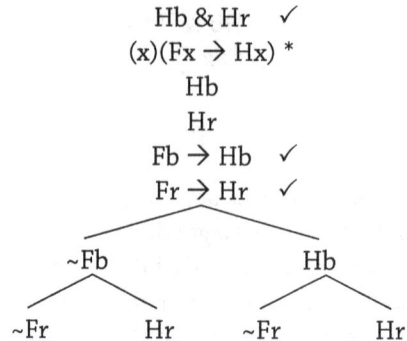

Since we created instances of the universal quantifier using all the names on the tree, and since the tree terminates with at least one open branch, then the two statements are consistent with each other.

The next pair of sentences, "If someone is free, then someone is happy" and "No one is happy" are symbolized and set up in a tree as follows

$$(\exists x)Fx \rightarrow (\exists y)Hy$$
$$\sim(\exists x)Hx$$

The first statement is a *conditional* whose antecedent and consequent are both existential statements, all of which you now know how to break down. However, the second statement is new for you as it's a negated existential statement. Fortunately, these are relatively easy. When you look at the original statement, "No one is happy" you may remember (from Chapter Four) that an equally acceptable way of symbolizing it is (x)~Hx. This would read, "Everyone is unhappy" (or: For any person, they won't be happy), which is just another way of saying that no one is happy. The two ways of symbolizing the statement are equivalent. Now when you symbolize it this second way the main connective becomes the universal quantifier, and you already have a rule for breaking down these kinds of statements. The same thing works for changing negated universal statements into existential statements.

*Thus, the rule for breaking down negated quantified statements is to begin by switching the quantifier to its counterpart. So, if the quantifier is an existential, replace it with a universal; if it's a universal, replace it with an existential. Then move the tilde to the immediate right of the new quantifier, and check off the statement.*

~(x)Px✓   ~(∃x)Px✓
(∃x)~Px   (x)~Px

You now have enough rules to be able to complete the tree and determine if the two statements are consistent.

# CHAPTER TEN: REINTRODUCING NAMES, PREDICATES, QUANTIFIERS, AND IDENTITY

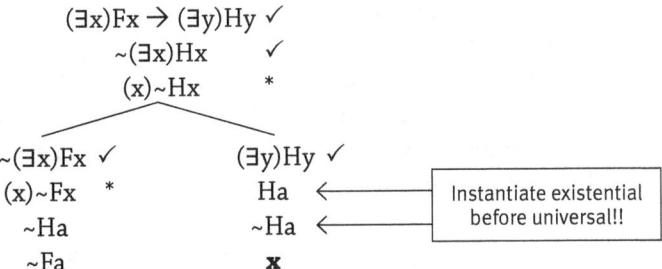

Notice the strategy employed on the right branch. We have to create an instance from both a universal and from an existential. We *create the existential instance first*, because any name we choose must be new to the tree. If we'd done the universal first, then we'd have to use another name for the existential, because that name would have already appeared on the tree—and the new name wouldn't give us a contradiction, so we'd need to instantiate the universal again to another new name. While technically this is okay, it does make for more work and longer trees when you don't need them.

For this tree, while the right branch closes, the left one remains open, thereby showing that there is an interpretation which makes both statements come out true. So these statements are *consistent*.

The final pair of statements from Example Set 10b:

"Someone is not free and not happy" and "Everyone is happy."

When these are symbolized and set up for a tree you have:

$$(\exists x)(\sim Fx \, \& \sim Hx) \quad \checkmark$$
$$(x)Hx \, *$$
$$\sim Fa \, \& \sim Ha \quad \checkmark$$
$$\sim Fa$$
$$\sim Ha$$
$$Ha$$
$$\mathbf{x}$$

Again, we instantiate existential before universal quantifications, because this strategy saves time and space. We see immediately that we can generate a contradiction if we instantiate the universal with the name 'a' to get Ha. And the tree closes.

Since the tree closes, we've shown that there is no interpretation which makes the two statements come out true at the same time, and so they are *not* consistent. However, while we now know they are not consistent, we don't yet know if they are contradictories or contraries. This time, since we've already run a consistency branch, instead of setting up a "biconditional" tree, just set up an inconsistency branch as a separate tree. We'll negate both statements and then run the tree to see if all branches close. If the tree closes, then you've shown that they are contradictories. If the tree terminates with an open branch, then you've shown that they are contraries.

$$
\begin{array}{c}
\sim(\exists x)(\sim Fx \mathbin{\&} \sim Hx) \quad \checkmark \\
\sim(x)Hx \quad \checkmark \\
(x)\sim(\sim Fx \mathbin{\&} \sim Hx) \quad * \\
(\exists x)\sim Hx \quad \checkmark \\
\sim Ha \\
\sim(\sim Fa \mathbin{\&} \sim Ha) \quad \checkmark \\
\diagup \quad \diagdown \\
\sim\sim Fa \ \checkmark \qquad \sim\sim Ha \ \checkmark \\
Fa \qquad\qquad Ha \\
\qquad\qquad\quad \mathbf{x}
\end{array}
$$

The left branch is open and when we double-check our starred universal statement, we see that there are no new statements we could derive that would lead to a contradiction on the left branch. Hence, we're done with the tree and we've shown that the two original statements are *contraries*.

## Exercise 10b

Symbolize the following pairs of statements, and then run a tree(s) on them to determine if they are consistent, contradictories, or contraries.

### TRANSLATION SCHEME
RD: People

| l: Lucio | a: Anne | j: Joanne | Hx: x is honest |
|---|---|---|---|
| Sx: x is a saint | Ex: x is an ethical egoist | Gx: x is selfish | |

1. All ethical egoists are selfish people. No selfish person is an ethical egoist.
2. There are some honest selfish people. There are no non-honest people who are not selfish.
3. If Lucio is a saint, then Anne is one too. Anne is a saint only if Lucio is.
4. Joanne is both an ethical egoist and a saint. All saints are non-ethical egoists.

5. Either Lucio is selfish or he's not. Lucio is not selfish, yet he's no saint.
6. Joanne is a saint if and only if she's an ethical egoist. Everyone who is not an ethical egoist is a saint.
7. Some saints are non-selfish people. No non-selfish people are ethical egoists, yet some saints are ethical egoists.
8. A selfish person is an ethical egoist. Joanne is an ethical egoist but not selfish.
9. Saints are neither selfish nor ethical egoists. However, some selfish people are saints, though not ethical egoists.
10. Everyone is either a saint, an ethical egoist, or selfish. Lucio is none of these.

Use trees to determine if the following pairs are consistent, contradictories, or contraries. Then translate the statements into English language sentences using the above translation scheme.

11. (∃x)(~Sx & ~Ex), (y)(Ey → Sy)
12. Sl & ~El, (∃z)(Sz & ~Ez)
13. (y)[(Sy & ~Ey) → ~Gy], Sj & (~Ej & Gj)
14. (z)(Gz & Ez), Sj & ~Gj
15. (Ga → Ea) & ~Sa, (x)(Ex → ~Sx)
16. (∃w)(Sw v Ew), ~(∃x)(Sx & Ex)
17. (z)[Sz v (Ez v Gz)], ~[(Sl v El) v Gl]
18. Ga & Sa, ~(∃x)(Gx & Sx)
19. (∃x)~(Sx & Gx), Sj & Gj
20. (y)[(Ey & Sy) → ~Ey], (∃z)(Sz & ~Ez)

## IMPLICATION, EQUIVALENCE, AND IDENTITY

Throughout this chapter we have been reintroducing the full LOLA that we first presented in Chapters Two through Six, with its individual constants, predicates, quantifiers, etc. At each step we have also revisited certain logical concepts that you initially encountered in Chapters Seven through Eight: logical truth, logical falsehood, contingency, consistency. For the last part of this chapter we'll reintroduce the process of symbolizing identity statements and incorporate the logical concepts of implication and equivalence.

~a = a
X
a=b
~a=b
X

Recall that an identity statement functions like a two place predicate. But instead of two distinct things, identity tells us that two names or bound variables are naming the same thing. Because of this, we need to introduce two new rules.

First, given what we just said, you can see that *if the main connective of the formula you're considering is an identity relationship, then you can't break it down any further.*

One difference, though, involves the nature of certain negated identity relationships, ones like the following:

1. ~o=o
2. ~t=t
3. ~d=d

*These statements are all asserting something that is impossible*, namely that a given thing is *not* identical to itself. These statements are just as much self-contradictions as the conjunction of a statement and its negation. This gives us a new rule for truth trees. *Any negated identity statement that has the same letter for both components will function in a truth tree the way that any contradiction would; it will close the branch that it appears on.*

The second rule involves a more detailed discussion of the nature of identity statements. We'll explain it by reintroducing the notion of implication.

### Example Set 10c

**TRANSLATION SCHEME**
RD: Positive numbers

| o: one | t: two | u: four | f: five | d: ten | w: twenty |
|---|---|---|---|---|---|
| Ex: x is even | Gxy: x is greater than y | Lxy: x is less than y | Pxy: x is added to y | Sxy: x is subtracted from y | Bxyz: x is between y and z |

Recall from the previous chapter that to say one statement *implies* another means that *it's impossible for the first statement to be true and the second statement to be false*. Consider the following pair of statements; we want to determine whether the first statement implies the second.

1. (∃x)(Gxo & x=f), Gfo (o: one; f: four; Gxy: x is greater than y)

To do this you set up a tree with the first statement on top and then the negation of the second statement below it. If the tree terminates with *all closed branches* then you've demonstrated that the first statement *does* imply the second.

$$(\exists x)(Gxo \text{ \& } x=f) \checkmark$$
$$\sim Gfo$$

As the main operator is the existential quantifier, you must use the rule governing existential statements in truth trees. Since we have to pick an individual constant to replace all the x's that are bound to the quantifier, and we can't pick either "o" or "f," we'll select "m."

We then break the statement down according to the rules governing conjunctions.

$$(\exists x)(Gxo \ \& \ x=f) \ \checkmark$$
$$\sim Gfo$$
$$Gmo \ \& \ m=f \quad \checkmark$$
$$Gmo$$
$$m=f$$

Now if we stopped here it would seem as there is no implication relationship because the tree terminates with an open branch. However, recall what we said earlier when we introduced the rule governing universal statements and truth trees. The goal of constructing a tree is to try to get it to terminate with all closed branches, making sure you always follow the rules. Therefore, you're always on the lookout for a way to close down a branch, and you've not completed the branch if there *is* still something to be done that can close it. In the tree above, we have exactly that situation, and we will now provide you with the rule that will let you accomplish this.

There are two different predicate statements standing alone, ~Gfo and Gmo. There is also an identity claim standing alone, m=f. What the identity statement says is that m is identical to f, which means that they are completely interchangeable. Wherever you have an f, you could replace it with an m. Wherever you have an m, you could replace it with an f. So, *any predicated statement or negated predicated statement along a branch containing the identity statement can have an individual constant replaced by its identical counterpart.*

This leads to the second rule governing identity statements and trees. If you have a predicate statement or the negation of a predicate statement and an identity statement, do the following: Look to see if either name in the identity statement is used in the predicate statement or the negated predicate statement. If it is, create a new instance of the predicate statement or negated predicate statement that substitutes the name that was there, with the other name from the identity claim. When you've completed all that, put a star by the identity statement so that you know that you may have to come back and repeat the process if a new predicate statement shows up on that branch.

In the tree above this would result in the following,

$(\exists x)(Gxo \,\&\, x=f)$ ✓
~Gfo
Gmo & m=f   ✓
Gmo
m=f *
~Gmo
x

We applied the rule using the identity statement and the negated predicate statement, ~Gfo. Since m is identical to f, then we can replace the f with an m and create a new version of the statement which reads, ~Gmo. Once we've done this we see that there is a contradiction on the branch and the tree terminates with no open branches. We could just as easily have taken the statement Gmo and substituted an f for the m in it, thereby giving us Gfo. Notice that this also would have resulted in a contradiction because you have Gfo and ~Gfo.

To determine whether the above two statements are equivalent we need to see if they mutually imply each other. We've already found that the first statement, $(\exists x)(Gxo \,\&\, x=f)$, implies the second, Gfo and if it turns out that this works both ways, then we'll have shown they are equivalent statements. This time we put Gfo on top and the negation of $(\exists x)(Gxo \,\&\, x=f)$ underneath it.

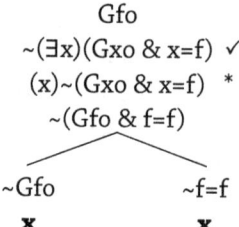

Gfo
~$(\exists x)(Gxo \,\&\, x=f)$  ✓
$(x)$~$(Gxo \,\&\, x=f)$   *
~(Gfo & f=f)

~Gfo         ~f=f
x             x

Since this tree also closes, we know *that $(\exists x)(Gxo \,\&\, x=f)$ and Gfo mutually imply each other and hence they are equivalent.* As you were working through the tree above, you may have thought about using o instead of f to instantiate the universal statement $(x)$~$(Gxo \,\&\, x=f)$. If you had, you would have created the following tree,

Gfo
~$(\exists x)(Gxo \,\&\, x=f)$  ✓
$(x)$~$(Gxo \,\&\, x=f)$   *
~(Goo & o=f)   ✓

~Goo          ~o=f

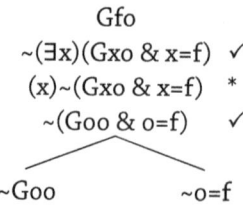

Since there are no contradictions, this might have led you to the mistaken idea that the tree terminates with open branches. However, recall what we said when we introduced the rule for universal statements. Your goal is to try to close the branches and if going back to the universal statement and instantiating to a new individual constant will close a branch then you must do so.

Since you still have another letter that you could use, namely f, you must go back to the universal statement and create new instances under both your branches. This will generate the following tree,

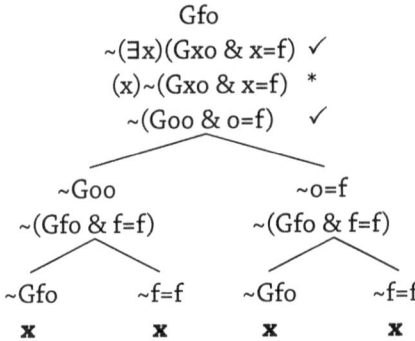

The tree eventually terminates with all closed branches, but it's a larger, more cumbersome tree. This doesn't make it wrong in any way, but it does illustrate that if you use some foresight and planning you can make it much easier on yourself while you're making choices about what to derive.

## Exercise 10c

For each pair of statements below, construct trees to see if the 1st statement implies the 2nd. If it does, construct a second tree to determine if the two statements are equivalent.

1. Pb & (x)(Px → x=b), ~~Pb & (x)(~x=b → ~Px)
2. Pb & (x)(Px → x=b), ~[~Pb v ~(x)(Px → x=b)]
3. (∃x)Mx → (x)Hx, (∃x)Mx → Hm
4. (x)(~x=h → Fhx), ~(∃x)~(x=h v Fhx)
5. ~(~a=b v b=a), Mba
6. (∃x)[Px & (y)(Py → x=y)], ~(Pa & Pb)
7. ~Ms & (x)(~x=s → Mx), ~[(∃x)(~x=s & ~Mx) v Ms]
8. (x)[(Dx & ~x=h) → Fhx], Db & (~Fbh & ~b=h)
9. a=b & Eb, Ea
10. p=s → (x)(Px & Ex), ~[p=s & ~(x)(Px & Ex)]

# Unit Two Review

## TRUTH-FUNCTIONAL DEFINITIONS: BASIC TRUTH TABLES

| NEGATION | |
|---|---|
| A | ~A |
| T | F |
| F | T |

| CONJUNCTION | | |
|---|---|---|
| A | B | A & B |
| T | T | T |
| T | F | F |
| F | T | F |
| F | F | F |

| DISJUNCTION | | |
|---|---|---|
| A | B | A v B |
| T | T | T |
| T | F | T |
| F | T | T |
| F | F | F |

| CONDITIONAL | | |
|---|---|---|
| A | B | A → B |
| T | T | T |
| T | F | F |
| F | T | T |
| F | F | T |

| BICONDITIONAL | | |
|---|---|---|
| A | B | A ↔ B |
| T | T | T |
| T | F | F |
| F | T | F |
| F | F | T |

## TRUTH TABLE SET UP

Truth input rows: All possible combinations of truth assignments in $2^n$ rows

| Statement letters | | Compound Statement(s) | |
|---|---|---|---|
| A | B | A → B | A ↔ B |
| T | T | | |
| T | F | | |
| F | T | | |
| F | F | | |

## BASIC TREES

**Conjunctions**
A & B ✓
A
B

**~Conjunctions**
~(A & B) ✓
　　　／＼
　~A　　~B

**Disjunctions**
A v B ✓
　／＼
A　　B

**~Disjunctions**
~(A v B) ✓
~A
~B

**Double ~**
~~A ✓
A

**Conditional**
A → B ✓
　／＼
~A　　B

**~Conditional**
~(A → B) ✓
A
~B

**Biconditional**
A ↔ B ✓
　／＼
A　　~A
B　　~B

**~Biconditional**
~(A ↔ B) ✓
　／＼
A　　~A
~B　　B

## QUANTIFIER AND IDENTITY TREE RULES

**(∃x) Instance**
(∃x)Px ✓
Pa
('a' new to tree)

**(x) Instance**
(x)Px *
Pa
(any name)

**(∃x) Negation**
~(∃x)Px ✓
(x)~Px

**(x) Negation**
~(x)Px ✓
(∃x)~Px

**ID Close**
n=a
~n=a
x

**ID Switch**
Pab
~Pan
n=b *
~Pab
x

**ID Close**
~n=n
x

## LOGICAL PROPERTIES

*L-Falsehood*

Df: Statement that is unconditionally false.
Table: A statement that has all Fs under its main connective.
Tree: Statement's tree has all closed branches.

```
(A & B) & ~(A v B)  ✓
    A & B           ✓
    ~(A v B)        ✓
      A
      B
      ~A
      ~B
      x
```

| A | B | (A & B) & ~ (A v B) |
|---|---|---|
| T | T | T **F** F T |
| T | F | F **F** F T |
| F | T | F **F** F T |
| F | F | F **F** T F |

*L-Truth*

Df: Statement that is unconditionally true.
Table: A statement that has all Ts under its main connective.
Tree: A statement whose *negation* yields all closed branches.

```
~[(A v B) v ~(A v B)]  ✓
    ~(A v B)           ✓
    ~~(A v B)          ✓
      A v B
      ~A
      ~B
     /    \
    A      B
    x      x
```

| A | B | (A v B) v ~ (A v B) |
|---|---|---|
| T | T | T **T** F T |
| T | F | T **T** F T |
| F | T | T **T** F T |
| F | F | F **T** T F |

*Contingent*

Df: Statement that is true under some conditions and false under some conditions.
Table: A statement that has at least one T and at least one F under its main connective.
Tree: Statement's tree, and tree of denial of statement, both have at least one open branch.

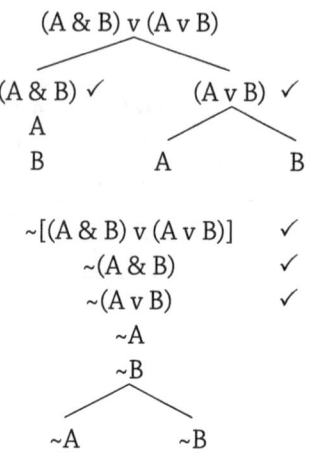

| A | B | (A & B) v (A v B) |
|---|---|---|
| T | T | T **T** T |
| T | F | F **T** T |
| F | T | F **T** T |
| F | F | F **F** F |

## LOGICAL RELATIONS

*Consistent*

Df: Possible for statements to be true at the same time.
Table: At least 1 row shows true for each statement's final value.
Tree: Tree with both statements will have at least 1 open branch.

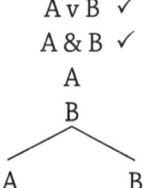

| A | B | A v B | A & B |
|---|---|---|---|
| T | T | T | T |
| T | F | T | F |
| F | T | T | F |
| F | F | F | F |

*Contrary*

Df: *Inconsistent* statements that can both be false.
Table: No row shows both statements true; at least one row shows both false.
Tree: *For statements you know are inconsistent,* a tree that negates both statements will have at least one open branch.

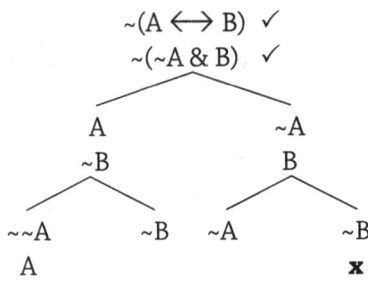

| A | B | A ⟷ B | A & B |
|---|---|---|---|
| T | T | T | T |
| T | F | T | F |
| F | T | T | F |
| F | F | F | F |

*Contradiction*

Df: Statements always have opposite truth values.
Table: Every row shows opposite truth values for the statements.
Tree: On a tree whose first statement is a biconditional of the 2 statements, every branch will close.

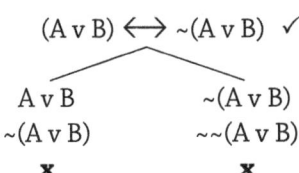

| A | B | A v B | ~(A v B) |
|---|---|---|---|
| T | T | T | F |
| T | F | T | F |
| F | T | T | F |
| F | F | F | T |

*Implication* ∴

Df: 1st statement ∴ 2nd if it's *impossible* for the 1st to be true and the 2nd false.
Table: No row shows the 1st ~(A v B) statement true and the 2nd ~A & ~B false.
Tree: Tree of the 1st statement ~(A v B) and the denial of the 2nd ~(~A & ~B) yields all closed branches.

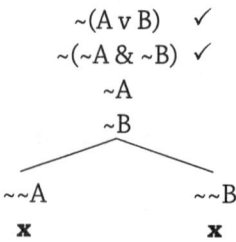

| A | B | ~(A v B) | ~A & ~B |
|---|---|----------|---------|
| T | T | F | F |
| T | F | F | F |
| F | T | F | F |
| F | F | T | T |

*Equivalence*

Df: Statements that always have the same truth value; statements that mutually imply each other.
Table: All rows show the same truth values (see previous example).
Tree: Construct two implication trees: 1st statement and denial of 2nd AND 2nd statement and denial of 1st. Both trees will have all closed branches.

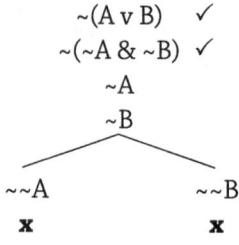
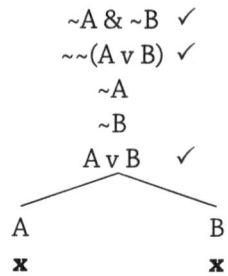

# Unit Two: Answers to Selected Problems

**CHAPTER SEVEN**

### Exercise 7a

5. John doesn't like cats or dogs. (D: John likes dogs; C: John likes cats.) **~(D v C) OR ~D & ~C**
8. Lorie won't graduate this term unless she finishes her coursework. (G: Lorie graduates this term; F: Lorie finishes her coursework.) **L → F**
9. One will have a fire if and only if there is fuel, oxygen, and an ignition of some sort. (F: There is a fire; S: There is fuel; O: There is oxygen; I: There is ignition of some sort.) **F ⟷ [(S & O) & I]**
11. The theater will be crowded tonight but safe, if the security people know what they're doing. (C: The theater is crowded tonight; S: The theater is safe; K: The security people know what they're doing.) **C & (K → S) OR K → (C & S)**
15. John's having a severe medical condition is a sufficient condition for his being discharged from the army. (M: John has a severe medical condition; D: John is discharged from the army.) **M → D**

### Exercise 7b

2. M ⟷ ~L **Mary will come if and only if it's not late.**
4. (T v L) & ~(T & L) **We'll be on time or it's late, but not both.**
5. E → C **If Evan wants to join us, we can meet you at 10:00.**
7. J → (M & ~E) **If John joins us, then Mary will come but Even won't want to join us.**
9. C & [A & {W → (L → ~T)}] **Although we can meet you at 10:00, all of us are busy and, if we wait for Sydney, then if it's late, we won't be on time.**

## Exercise 7c

1. A & (~B v C)

| A | B | C | A & (~B v C) |
|---|---|---|---|
| T | T | T | T F T |
| T | T | F | F F F |
| T | F | T | T T T |
| T | F | F | T T T |
| F | T | T | F F T |
| F | T | F | F F F |
| F | F | T | F T T |
| F | F | F | F T T |

2. C ↔ (D & C)

| C | D | C ↔ (D & C) |
|---|---|---|
| T | T | T T |
| T | F | F F |
| F | T | T F |
| F | F | T F |

10. K v (K → ~K)

| K | K v (K → ~K) |
|---|---|
| T | T F F |
| F | T T T |

17. There's a problem with the carburetor, although if Morton has to spend much money on it, he'll just scrap the car and buy a used one. (C: There's a problem with the carburetor; M: Morton has to spend much money on the car; S: Morton scraps the car; U: Morton buys a used car.)
    **C & [M → (S & U)]**

| C | M | S | U | C & [M → (S & U)] | | |
|---|---|---|---|---|---|---|
| T | T | T | T | T | T | T |
| T | T | T | F | F | F | F |
| T | T | F | T | F | F | F |
| T | T | F | F | F | F | F |
| T | F | T | T | T | T | T |
| T | F | T | F | T | T | F |
| T | F | F | T | T | T | F |
| T | F | F | F | T | T | F |
| F | T | T | T | F | T | T |
| F | T | T | F | F | F | F |
| F | T | F | T | F | F | F |
| F | T | F | F | F | F | F |
| F | F | T | T | F | T | T |
| F | F | T | F | F | T | F |
| F | F | F | T | F | T | F |
| F | F | F | F | F | T | F |

18. The lamps go on automatically at dusk, and then, if there's not enough light to see, you can turn on the rest of them manually. (A: The lamps go on automatically; E: There's enough light to see; T: A person can turn on the rest of the lights manually.)    **A & (~E → T)**

| A | E | T | A & (~E → T) | | |
|---|---|---|---|---|---|
| T | T | T | T | F | T |
| T | T | F | T | F | T |
| T | F | T | T | T | T |
| T | F | F | F | T | F |
| F | T | T | F | F | T |
| F | T | F | F | F | T |
| F | F | T | F | T | T |
| F | F | F | F | T | F |

**Exercise 7d**

1. (D & E) v ~(D & E)

| D | E | (D & E) v ~(D & E) |
|---|---|---|
| T | T | T  T F  T |
| T | F | F  T T  F |
| F | T | F  T T  F |
| F | F | F  T T  F |

**ALL Ts IN FINAL COLUMN: L-TRUE**

2. (G v ~G) → (B & ~B)

| G | B | (G v ~ G) → (B & ~B) |
|---|---|---|
| T | T | T F  F  F F |
| T | F | T F  F  F T |
| F | T | T T  F  F F |
| F | F | T T  F  F T |

**ALL Fs IN FINAL COLUMN: L-FALSE**

7. ~[(Q → S) & ~Q]

| Q | S | ~[(Q → S) & ~Q] |
|---|---|---|
| T | T | T  T  F F |
| T | F | T  F  F F |
| F | T | F  T  T T |
| F | F | F  T  T T |

**AT LEAST 1 T AND 1 F IN FINAL COLUMN: CONTINGENT**

**Exercise 7e**

3. An aardvark is not a car nor is it a person, nor a whale.
   **(x){Ax → [~Cx & (~Px & ~Wx)]}**
4. Lane is not a singer, but everyone likes singing to Pam.
   **~Vl & (x)(Px → Lxp)**
6. People are not faster than cars, but whales are if they don't quit.
   **(x)(y)[(Px & Cy) → ~Fxy] & (x)(y)[(Wx & Cy) → (~Qx → Fxy)]**
9. If an aardvark is faster than a fish, then it is speedy.
   **(x)(y){[Ax & (Fy & Fxy)] → Sx}**
12. A person quits being a singer only if they don't like singing to somebody.
    **(x)[(Px & ~Vx) → ~(∃y)(Py & Lxy)]**

# CHAPTER EIGHT

### Exercise 8a

1. Tim is a bore, yet he's been promoted over Susan who's not.
   **(Bt & Pts) & ~Bs**
2. A necessary condition for being a fireman is being a hard worker.
   **(x)(Fx → Hx)**
6. There are a number of hardworking lawyers, Matt just isn't one of them.
   **(∃x)(Lx & Hx) & (Lm & ~Hm)**

### Exercise 8b

1. Either we go to the concert or the theater, but we don't just stay home. (C: We go to the concert; T: We go to the theater; H: We stay home.)    **(C v T) & ~H**

   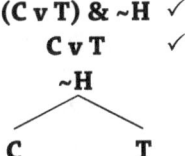

   **The statement is true when H is false and either C or T is true.**

4. It's false that both Bob and John cheated on the exam. (B: Bob cheated on the exam; J: John cheated on the exam.)    **~(B & J)**

   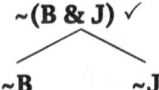

   **The statement is true when either B or J is false.**

5. It's not true that either you have to vote for a Democrat or a Republican. (D: You have to vote for a Democrat; R: You have to vote for a Republican.)  **~(D v R)**

$$\begin{array}{c} \sim(D \vee R) \checkmark \\ \sim D \\ \sim R \end{array}$$

**The statement is true when D and R are both false.**

6. Each of us wants to have a successful career, personal life, good health, and not hurt other people. (C: We want to have a successful career; P: We want to have a successful personal life; H: We want to have good health; O: We want to harm other people.)  **(C & P) & (H & ~O)**

$$\begin{array}{c} (C \& P) \& (H \& \sim O) \checkmark \\ C \& P \quad \checkmark \\ H \& \sim O \quad \checkmark \\ C \\ P \\ H \\ \sim O \end{array}$$

**The statement is true when C, P, and H are true and O is false.**

### Exercise 8c

1. You have fire just in case you have air, a fuel source, and a spark. (F: You have a fire; A: You have air; S: You have a fuel source; P: You have a spark.)  **F ⟷ [(A & S) & P]**

$$\begin{array}{c} F \leftrightarrow [(A \& S) \& P] \checkmark \end{array}$$

```
              F ⟷ [(A & S) & P] ✓
              ╱                ╲
            F                    ~F
        (A & S) & P ✓       ~[(A & S) & P] ✓
           A & S ✓              ╱        ╲
             P               ~(A & S) ✓    ~P
             A                 ╱    ╲
             S               ~A      ~S
```

**The statement is true when F, P, A, and S are all true OR when F is false and either P, A, or S is also false.**

6. It is not true that Sam gave a false deposition and that he is hiding something. (S: Sam gave a true deposition; H: Sam is hiding something.)
~(~S & H)

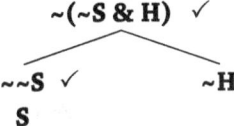

**The statement is true when S is true or H is false.**

8. Winning California, New York, Florida, and either Michigan or Pennsylvania is a necessary and sufficient condition for any Democrat to win the presidency. (C: A Democrat wins California; N: A Democrat wins New York; F: A Democrat wins Florida; M: A Democrat wins Michigan; P: A Democrat wins Pennsylvania; W: A Democrat wins the presidency.)
{[(C & N) & F] & (M v P)} ⟷ W

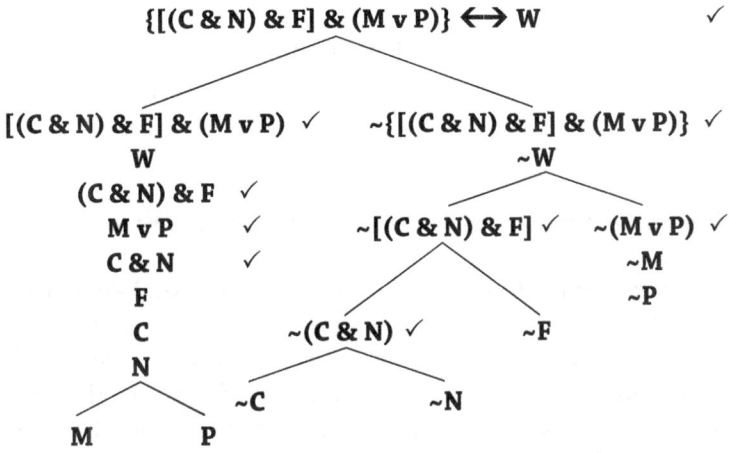

**The statement is true when W, F, C, and N are all true and either M or P is also true. OR When W is false and both M and P are false. OR When W is false and either F, C, or N is false.**

**Exercise 8d**

5. Assuming it doesn't rain, we can work on the project and make some progress. (R: It rains; P: We can work on the project; M: We make some progress.)  ~R → (P & M)

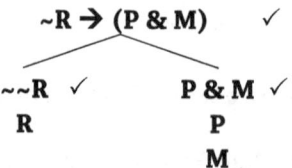

**The statement is true when either R is true or P and M are both true.**

7. It's not true that if I can't do this exercise I'm out of shape. (D: I can do this exercise; I: I'm in shape.)  ~(~D → ~I)

~(~D → ~I) ✓
~D
~~I  ✓
I

**The statement is true when D is false and I is true.**

9. Thomas Aquinas' five proofs for the existence of God work just in case you accept a particular understanding of the nature of causation, which I don't. (T: Thomas Aquinas' five proofs for the existence of God work; A: You accept a particular understanding of the nature of causation; I: I accept the particular understanding of the nature of causation.)
T ↔ (A & ~I)

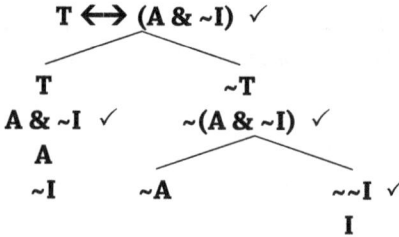

**Statement true when T and A are true and I is false OR when T is false and either A is false or I is true.**

12. (P → M) → (~P v M)

```
         (P → M) → (~P v M) ✓
          /              \
    ~(P → M) ✓         ~P v M ✓
        P                /    \
       ~M              ~P      M
```

**Open branches: Run a 2nd tree with the statement negated.**

```
    ~[(P → M) → (~P v M)] ✓
         P → M              ✓
         ~(~P v M)          ✓
          ~~P               ✓
          ~M
           P
          / \
        ~P   M
         x   x
```

**All branches close; this means the negated statement is L-false. Original statement is L-TRUE.**

13. (M & N) → ~(N & M)

```
           (M & N) → ~(N & M) ✓
          /                   \
     ~(M & N) ✓           ~(N & M) ✓
       /    \               /    \
     ~M     ~N            ~N     ~M
```

**Open branches: Run a 2nd tree with the statement negated.**

```
   ~[(M & N) → ~(N & M)] ✓
         M & N            ✓
         ~~(N & M)        ✓
         N & M            ✓
           M
           N
           N
           M
```

**Tree of negated statement has an open branch. Original statement is CONTINGENT.**

19. [M & (M → L)] & ~L

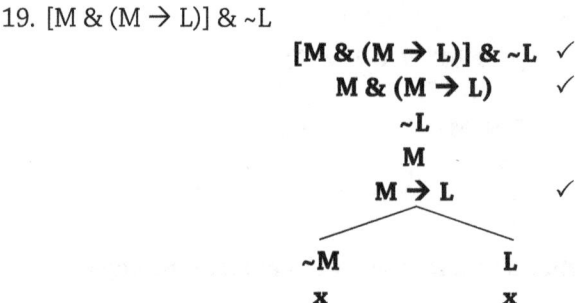

**All branches close: L-FALSE.**

20. (~Y & ~W) → ~(Y v W)

```
          (~Y & ~W) → ~(Y v W)  ✓
         ╱                    ╲
   ~(~Y & ~W) ✓            ~(Y v W) ✓
    ╱      ╲                  ~Y
  ~~Y ✓   ~~W ✓              ~W
   Y       W
```

**Open branches: Run a 2nd tree with the statement negated.**

```
     ~[(~Y & ~W) → ~(Y v W)]  ✓
            ~Y & ~W              ✓
            ~~(Y v W)            ✓
              ~Y
              ~W
             Y v W               ✓
            ╱    ╲
           Y      W
           x      x
```

**All branches close; this means the negated statement is L-false. Original statement is L-TRUE.**

21. ~[(~F v H) → (F → H)]

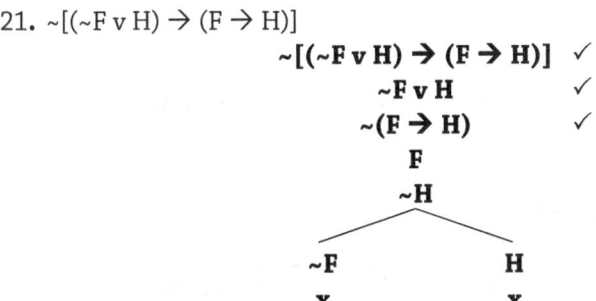

**All branches close: L-FALSE.**

22. ~[K & (K v O)]

~[K & (K v O)] ✓
```
        ~K         ~(K v O)✓
                     ~K
                     ~O
```

**Open branches: Run a 2nd tree with the statement negated.**

~~[K & (K v O)] ✓
K & (K v O) ✓
K
K v O
```
   K            O
```

**Tree of negated statement has open branches. Original statement is: CONTINGENT.**

### Exercise 8e (Translation Review)

6. Everyone insults someone, but not everyone fights a duel with someone.
   **(x)(∃y)[(Px & Py) → Ixy] & ~(x)(∃y)[(Px & Py) → Dxy]**
7. No one beat Jefferson in the election, though Adams did beat Burr.
   **~(∃x)(Px & Bxj) & Bab**
8. Adams did not kill Jefferson or anyone else for that matter.
   **~[Kaj v (∃x)(Px & Kax)]**
10. Jefferson and Adams were excellent shots, but in fact they did not duel with anybody. **(Ej & Ea) & (x)[Px → (~Djx & ~Dax)]**

# CHAPTER NINE

### Exercise 9a

Note: on those shown consistent by table, we've just shown the consistency branch of the tree.

1. ~(A & B), M → (B v ~M)

| A | B | M | ~(A & B) | M → (B v ~M) |
|---|---|---|----------|--------------|
| T | T | T | F    T   | T    T F     |
| T | T | F | F    T   | T    T T     |
| T | F | T | T    F   | F    F F     |
| T | F | F | T    F   | T    T T     |
| F | T | T | T    F   | T    T F     |
| F | T | F | T    F   | T    T T     |
| F | F | T | T    F   | F    F F     |
| F | F | F | T    F   | T    T T     |

**Consistent by table: Row 4 both statements can be true at the same time.**

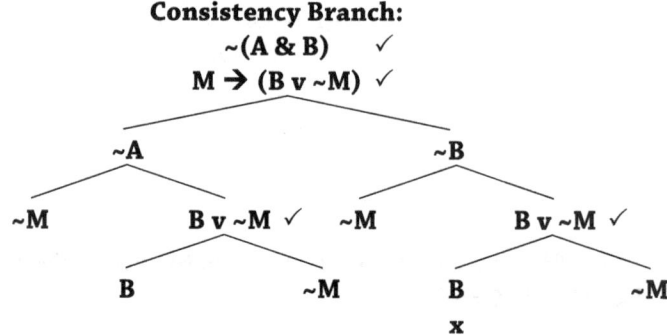

**Consistent by tree: Has at least one open branch.**

5. L & (Q v D), ~{(L & Q) v (L & D)}

| L | Q | D | L & (Q v D) | | ~{(L & Q) v (L & D)} | | | |
|---|---|---|---|---|---|---|---|---|
| T | T | T | T | T | F | T | T | T |
| T | T | F | T | T | F | T | T | F |
| T | F | T | T | T | F | F | T | T |
| T | F | F | F | F | T | F | F | F |
| F | T | T | F | T | T | F | F | F |
| F | T | F | F | T | T | F | F | F |
| F | F | T | F | T | T | F | F | F |
| F | F | F | F | F | T | F | F | F |

**Contradictory by table: always opposite truth values.**

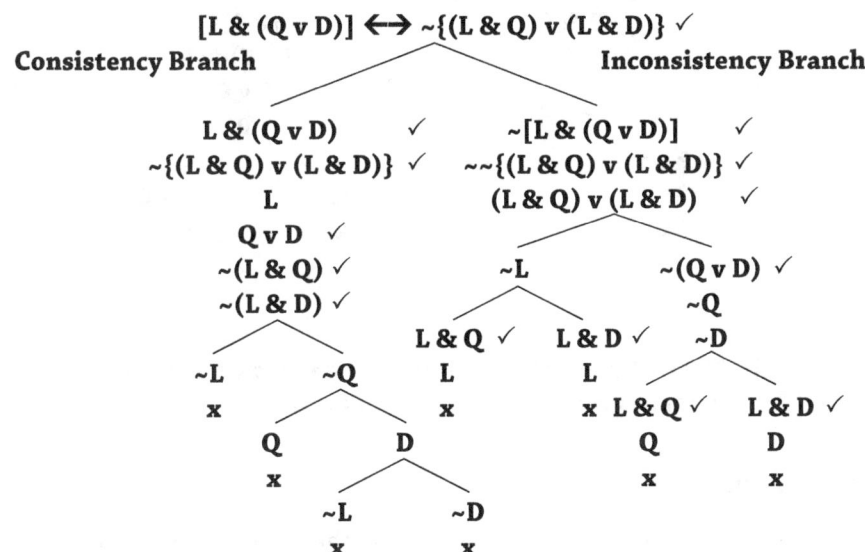

**Contradictory by tree: All branches close.**

9. ~[(P v C) & P], P & C

| P | C | ~[(P v C) & P] | P & C |
|---|---|---|---|
| T | T | F T T | T |
| T | F | F T T | F |
| F | T | T T F | F |
| F | F | T F F | F |

**Contrary by table: Can't both be true, can both be false.**

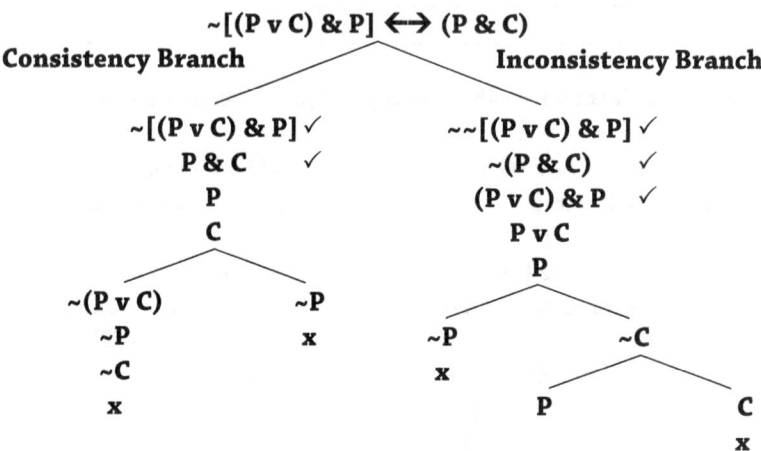

**CONTRARY by tree: Consistent branch all closed branches. Inconsistent branch with one open branch.**

### Exercise 9b

1. The Hope diamond is worth more than the McLeod quartz, but no one can afford the Hope diamond. **Wdq & ~(∃x)(Px & Axd)**
4. If no one has heard of the Jackson flintrock, then it's worthless.
   **~(∃x)(Px & Hxf) → Wf**
7. The McLeod quartz makes no one go mad. **(x)(Px → ~Cqx)**
10. Nothing is worth more than the Star of India unless it causes everyone to go mad. **(x)(y)[(Wxs & Py) → Cxy]**

## Exercise 9c

1. M & (B v C), C → ~B

| M | B | C | M & (B v C) | C → ~B |
|---|---|---|---|---|
| T | T | T | T  T | F  F |
| T | T | F | T  T | T  F |
| T | F | T | T  T | T  T |
| T | F | F | F  F | T  T |
| F | T | T | F  T | F  F |
| F | T | F | F  T | T  F |
| F | F | T | F  T | T  T |
| F | F | F | F  F | T  T |

**1st does *not* ∴ 2nd: Row 1 shows 1st T and 2nd F. (Rows 2 and 3 show 1st and 2nd are consistent.)**

**Implication Tree: 1st statement and denial of the 2nd:**

$$
\begin{array}{c}
M \,\&\, (B \vee C) \;\checkmark \\
\sim(C \rightarrow \sim B) \;\checkmark \\
M \\
B \vee C \\
C \\
\sim\sim B \;\checkmark \\
B \\
\diagup \;\;\; \diagdown \\
B \quad\quad\quad C
\end{array}
$$

**Open branches show it is possible for the 1st to be true and the 2nd false, so 1st does not imply the 2nd.**

**[Consistency Branch]**

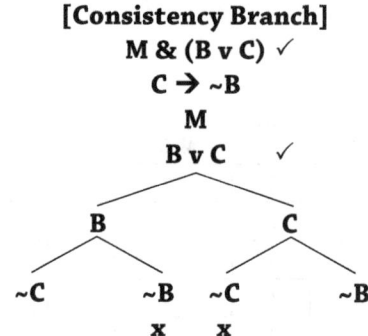

At least one open branch—possible both statements true at same time. **CONSISTENT STATEMENTS.**

8. P → ~Q, ~(Q → ~P)

| P | Q | P → ~Q | ~(Q → ~P) |
|---|---|--------|-----------|
| T | T | F F    | T   F F   |
| T | F | T T    | F   T F   |
| F | T | T F    | F   T T   |
| F | F | T T    | F   T T   |

**1st does *not* ∴ 2nd:** Rows 2, 3, 4 show 1st true and 2nd false.

Implication Tree: 1st statement and denial of the 2nd:

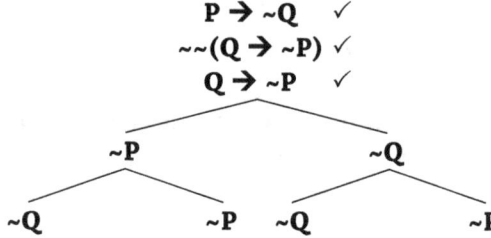

Open branches show it is possible for the 1st to be true and the 2nd false, so 1st does not ∴ the 2nd.

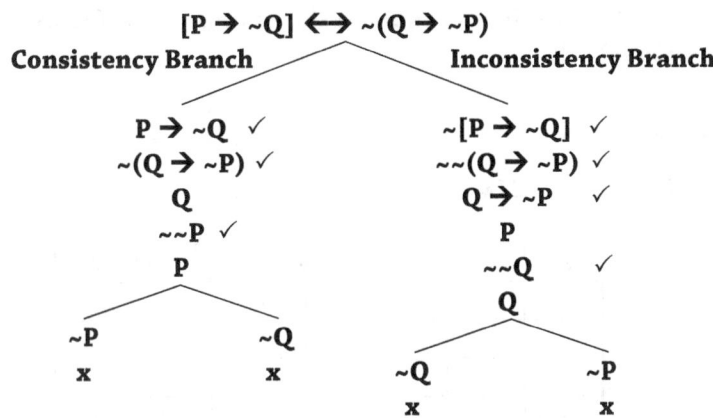

**CONTRADICTORY statements—can't have the same truth-value at the same time.**

10. F v (W v P), (F v W) v P

| F | W | P | F v (W v P) | | (F v W) v P | |
|---|---|---|---|---|---|---|
| T | T | T | T | T | T | T |
| T | T | F | T | T | T | T |
| T | F | T | T | T | T | T |
| T | F | F | T | F | T | T |
| F | T | T | T | T | T | T |
| F | T | F | T | T | T | T |
| F | F | T | T | T | F | T |
| F | F | F | F | F | F | F |

**1st ∴ 2nd because whenever the 1st is true, the 2nd is true.**

**Implication Tree: 1st and denial of 2nd:**

$$F \lor (W \lor P) \;\checkmark$$
$$\sim[(F \lor W) \lor P] \;\checkmark$$
$$\sim(F \lor W) \;\checkmark$$
$$\sim P$$
$$\sim F$$
$$\sim W$$

```
      F          W v P ✓
      x         /     \
                W      P
                x      x
```

**1st ∴ 2nd: all closed branches shows impossible for 1st to be true and 2nd false.**

**Exercise 9d**

2. (M v Z) v J, J v (M v Z)

| M | Z | J | (M v Z) v J | J v (M v Z) |
|---|---|---|---|---|
| T | T | T | T T | T T |
| T | T | F | T T | T T |
| T | F | T | T T | T T |
| T | F | F | T T | T T |
| F | T | T | T T | T T |
| F | T | F | T T | T T |
| F | F | T | F T | T F |
| F | F | F | F F | F F |

**1st ∴ 2nd: whenever 1st true, 2nd also true.
(Statements are equivalent because always have same truth value.)**

    (M v Z) v J  ✓
    ~[J v (M v Z)]  ✓
    ~J
    ~(M v Z)  ✓
    ~M
    ~Z

   M v Z ✓      J
                    x
M     Z
x     x

**1st ∴ 2nd: tree shows impossible for 1st to be true and 2nd false.**

J v (M v Z) ✓
~[(M v Z) v J] ✓
~(M v Z) ✓
~J
~M
~Z

```
        J              M v Z  ✓
        x
                    M        Z
                    x        x
```

**2nd ∴ 1st: tree shows impossible for 2nd to be true and 1st false. Mutual implication: EQUIVALENT.**

7. Q v R, ~(~R → Q)

| Q | R | Q v R | ~(~R → Q) |
|---|---|-------|-----------|
| T | T | T     | F F   T   |
| T | F | T     | F T   T   |
| F | T | T     | F F   T   |
| F | F | F     | T T   F   |

**1st does not imply 2nd, because 2nd can be false when 1st is true. (Contradictory statements because always have opposite truth values.)**

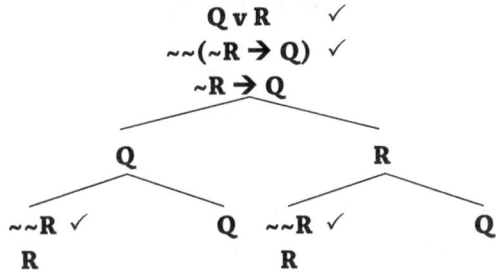

**1st does not imply 2nd because possible for 1st to be true and 2nd false.**

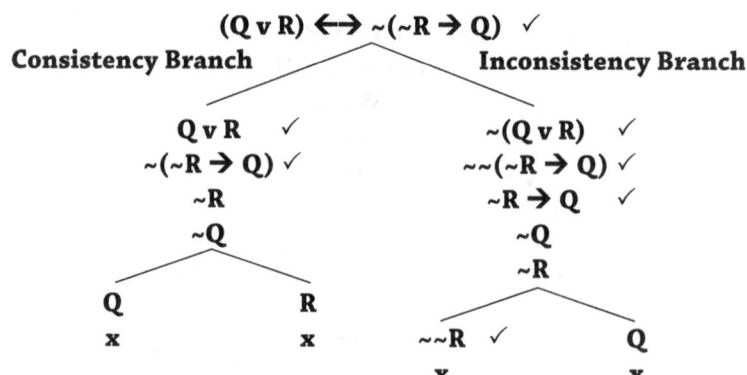

**CONTRADICTORY statements, they never have the same truth value.**

12. (D v E) & (D → E), E

| D | E | (D v E) & (D → E) | E |
|---|---|---|---|
| T | T | T T T | T |
| T | F | T F F | F |
| F | T | T T T | T |
| F | F | F F T | F |

**1st ∴ 2nd because whenever 1st is true, 2nd is also true. (Equivalent because always have the same truth value.)**

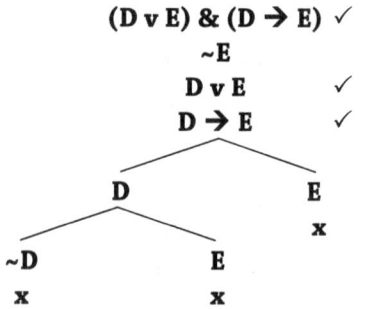

**1st ∴ 2nd because impossible for 1st to be true and 2nd false.**

# UNIT TWO: ANSWERS TO SELECTED PROBLEMS

$$E$$
$$\sim[(D \lor E) \mathbin{\&} (D \to E)] \checkmark$$

```
       ~(D v E) ✓        ~(D → E) ✓
         ~D                  D
         ~E                 ~E
          x                  x
```

**2nd ∴ 1st: impossible for 2nd to be true and 1st false.**
**Mutual Implication: EQUIVALENT.**

14. $J \leftrightarrow D$, $\sim(J \lor \sim D)$

| J | D | J ↔ D | ~(J v ~D) |
|---|---|-------|-----------|
| T | T | T     | F T F     |
| T | F | F     | F T T     |
| F | T | F     | T F F     |
| F | F | T     | F T T     |

**1st does not imply 2nd because 2nd can be false when 1st is true.**
**(Contrary statements: Both can't be true, but both can be false.)**

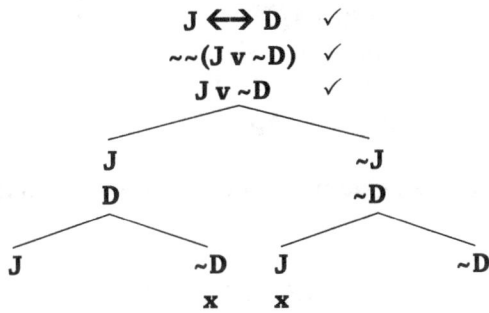

**1st does not imply 2nd because possible for 1st to be true and 2nd false.**

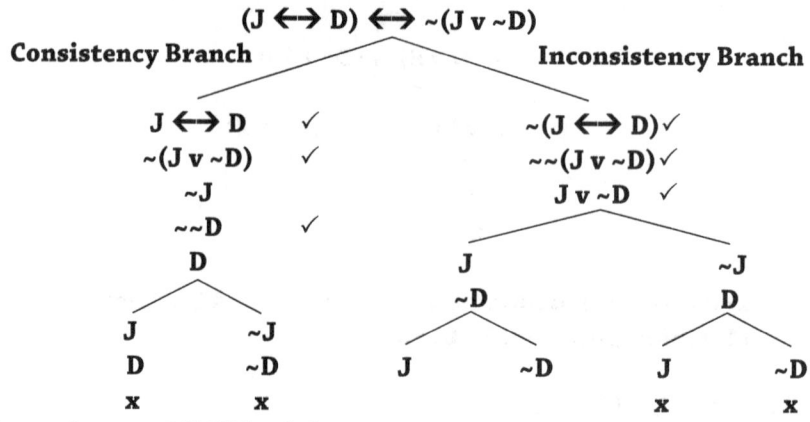

Inconsistent: CAN'T both be true.   Contrary: CAN both be false.

## CHAPTER TEN

### Exercise 10a

2. A sufficient condition for Plato being a Platonist is that he not be either a Sophist or an Aristotelian. **~(Sp v Ap) → Pp**

```
        ~(Sp v Ap) → Pp  ✓
           /        \
     ~~(Sp v Ap) ✓   Pp
      Sp v Ap ✓
       /    \
      Sp    Ap
```

**Not logically false; will run another tree to see if contingent or L-true.**

```
     ~[~(Sp v Ap) → Pp]  ✓
        ~(Sp v Ap)        ✓
           ~Pp
           ~Sp
           ~Ap
```

**The negated statement too is not L-false, so original statement is CONTINGENT.**

4. If Socrates is both a Sophist and not a Sophist, then he's also a Platonist.
   (Ss & ~Ss) → Ps

```
            (Ss & ~Ss) → Ps  ✓
           /              \
      ~(Ss & ~Ss) ✓        Ps
       /        \
     ~Ss        ~~Ss
                 Ss
```

**All open branches; not L-false. Run another tree to check if negation is L-false.**

$$\sim[(Ss \;\&\; \sim Ss) \to Ps] \;\; \checkmark$$
$$Ss \;\&\; \sim Ss \;\; \checkmark$$
$$\sim Ps$$
$$Ss$$
$$\sim Ss$$
$$x$$

**Negated statement is L-false, so original statement is L-TRUE.**

14. (Ap → ~Aa) & (Aa & Ap) **If Plato is an Aristotelian, then Aristotle isn't, but both Aristotle and Plato are Aristotelians.**

```
     (Ap → ~Aa) & (Aa & Ap)  ✓
            Ap → ~Aa
            Aa & Ap          ✓
               Aa
               Ap
              /  \
           ~Ap   ~Aa
            x     x
```

**All closed branches: statement is L-FALSE.**

15. Wpas → ~Waps **If Plato is more wrong than Aristotle about Socrates, then Aristotle can't be more wrong than Plato about Socrates.**

```
        Wpas → ~Waps  ✓
        ╱           ╲
     ~Wpas          ~Waps
```

**Open branches, not L-false. Run 2nd tree.**

```
   ~(Wpas → ~Waps)  ✓
         Wpas
        ~~Waps      ✓
         Waps
```

**Tree of negated original statement has open branch. Original statement is CONTINGENT.**

17. (Sp & ~Sp) → (Ps v ~Ps) **If Plato both is and isn't a Sophist, then Socrates either is or isn't a Platonist.**

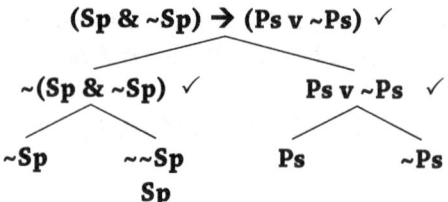

**All open branches. This tree says the statement is true whether Sp is true or false. It's true whether Ps is true or false—looks like it's L-true. But we'll run the 2nd tree to make sure.**

```
   ~[(Sp & ~Sp) → (Ps v ~Ps)]   ✓
            Sp & ~Sp             ✓
           ~(Ps v ~Ps)
               Sp
              ~Sp
               x
```

**All closed branches—negated statement is L-false; original statement is L-true. (A conditional whose antecedent is a L-false will always be L-true.)**

19. Ss v (Rsa & Rsp) **Either Socrates is a Sophist or he is more religious than both Aristotle and Plato.**

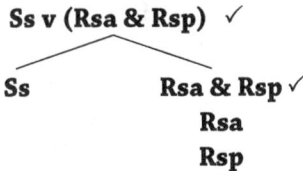

**Ss v (Rsa & Rsp)** ✓
    Ss      Rsa & Rsp ✓
                Rsa
                Rsp

**Open branches. Can you tell by looking if it is contingent or L-true? We'll run the 2nd tree.**

~[Ss v (Rsa & Rsp)] ✓
~Ss
~(Rsa & Rsp) ✓
~Rsa      ~Rsp

**Open branches. Not L-false; original statement is CONTINGENT.**

### Exercise 10b

4. Joanne is both an ethical egoist and a saint. All saints are non-ethical egoists. **Ej & Sj, (x)(Sx → ~Ex)**

Ej & Sj ✓
(x)(Sx → ~Ex) *
Ej
Sj
Sj → ~Ej ✓
~Sj      ~Ej
x         x

**Closed branches; not possible that both are true. INCONSISTENT. Check to see if both can be false.**

~(Ej & Sj)    ✓
~(x)(Sx → ~Ex)    ✓
(∃x)~(Sx → ~Ex)    ✓
~(Sa → ~Ea)    ✓
Sa
~~Ea    ✓
Ea

```
      ~Ej       ~Sj
```

**Open branches; possible for both to be false. CONTRARIES.**

6. Joanne is a saint if and only if she's an ethical egoist. Everyone who is not an ethical egoist is a saint. **Sj ↔ Ej, (x)(~Ex → Sx)**

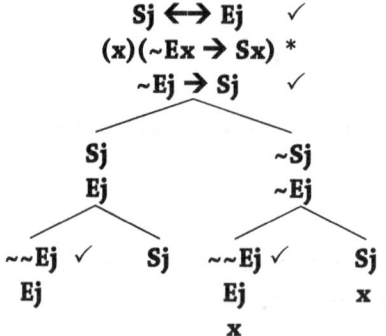

**Open branches—possible for both to be true. CONSISTENT.**

10. Everyone is either a saint, an ethical egoist, or selfish. Lucio is none of these. **(x)[Sx v (Ex v Gx)], ~Sl & (~El & ~Gl)**

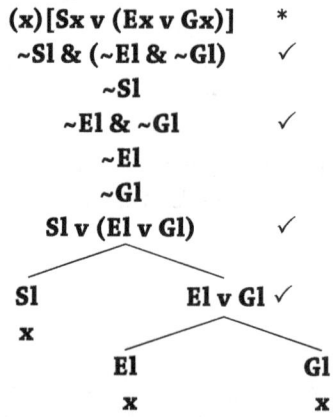

**Closed branches; not possible that both are true. INCONSISTENT. Check to see if both can be false.**

~(x)[Sx v (Ex v Gx)]   ✓
~[~Sl & (~El & ~Gl)]   ✓
(∃x)~[Sx v (Ex v Gx)]   ✓
~[Sa v (Ea v Ga)]   ✓
~Sa
~(Ea v Ga)   ✓
~Ea
~Ga

        ~~Sl ✓     ~(~El & ~Gl) ✓
        Sl
                   ~~El ✓  ~~Gl ✓
                   El       Gl

**Open branches; possible for both to be false. CONTRARIES.**

15. (Ga → Ea) & ~Sa, (x)(Ex → ~Sx) **If Anne is selfish, then she's an ethical egoist, and she's no saint. No ethical egoists are saints.**

(Ga → Ea) & ~Sa   ✓
(x)(Ex → ~Sx)   *
Ga → Ea   ✓
~Sa
Ea → ~Sa   ✓

      ~Ga            Ea
~Ea    Sa   ~Ea   ~Sa
         x

**Open branches—possible for both to be true. CONSISTENT.**

16. (∃w)(Sw v Ew), ~(∃x)(Sx & Ex) **Someone is either a saint or an ethical egoist. No one is both a saint and an ethical egoist.**

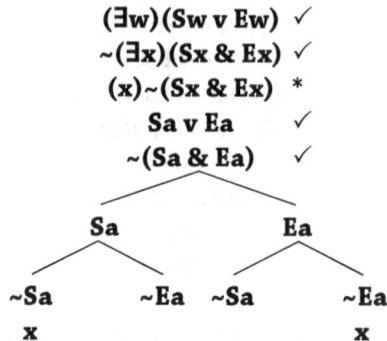

**Open branches—possible for both to be true. CONSISTENT.**

17. (z)[Sz v (Ez v Gz)], ~[(Sl v El) v Gl] **Everyone's either a saint, an ethical egoist, or selfish. It's false that Lucio is a saint, an ethical egoist, or selfish.**

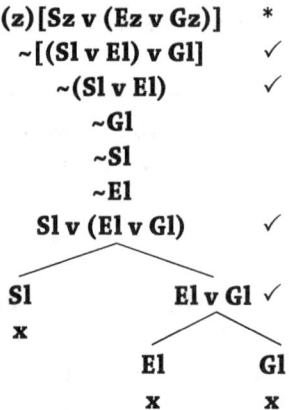

**Closed branches; impossible both are true. INCONSISTENT. Check to see if both can be false.**

~(z)[Sz v (Ez v Gz)] ✓
~~[(Sl v El) v Gl] ✓
(Sl v El) v Gl ✓
(∃z)~[Sz v (Ez v Gz)] ✓
~[Sa v (Ea v Ga)] ✓
~Sa
~(Ea v Ga) ✓
~Ea
~Ga

    Sl v El        Gl

Sl    El

**Open branches; possible for both to be false. CONTRARIES.**

20. (y)[(Ey & Sy) → ~Ey], (∃z)(Sz & ~Ez) **All ethical egoists who are saints are not ethical egoists. Some saints are not ethical egoists.**

(y)[(Ey & Sy) → ~Ey] *
(∃z)(Sz & ~Ez) ✓
Sa & ~Ea ✓
Sa
~Ea
(Ea & Sa) → ~Ea ✓

~(Ea & Sa) ✓    ~Ea

~Ea   ~Sa
     x

**Open branches—possible for both to be true. CONSISTENT.**

**Exercise 10c**

1. Pb & (x)(Px → x=b); ~~Pb & (x)(~x=b → ~Px)

          Pb & (x)(Px → x=b) ✓
     ~[~~Pb & (x)(~x=b → ~Px)] ✓
              Pb
         (x)(Px → x=b)    *

```
      ~~~Pb ✓          ~(x)(~x=b → ~Px) ✓
       ~Pb             (∃x)~(~x=b → ~Px) ✓
        x              ~(~a=b → ~Pa)    ✓
                        ~a=b
                        ~~Pa
                         Pa
                        Pa → a=b
                   ┌──────────┐
                  ~Pa         a=b
                   x           x
```

**1st ∴ 2nd:** tree of 1st and denial of 2nd results in all closed branches—showing it's impossible for the 1st to be true and the 2nd false.

      ~~Pb & (x)(~x=b → ~Px) ✓
      ~[Pb & (x)(Px → x=b)] ✓
           ~~Pb        ✓
            Pb
      (x)(~x=b → ~Px)    *

```
      ~Pb              ~(x)(Px → x=b) ✓
       x               (∃x)~(Px → x=b) ✓
                       ~(Pa → a=b)    ✓
                        Pa
                        ~a=b
                        ~a=b → ~Pa    ✓
                   ┌──────────┐
                 ~~a=b         ~Pa
                   x            x
```

**2nd ∴ 1st:** tree of 2nd and denial of the 1st results in closed branches. Mutual Implication: **EQUIVALENT**.

3. $(\exists x)Mx \to (x)Hx$; $(\exists x)Mx \to Hm$

$(\exists x)Mx \to (x)Hx$ ✓
$\sim[(\exists x)Mx \to Hm]$ ✓
$(\exists x)Mx$
$\sim Hm$

```
        ~(∃x)Mx            (x)Hx *
           x                Hm
                             x
```

**1st ∴ 2nd: tree of 1st and denial of 2nd results in all closed branches—showing it's impossible for the 1st to be true and the 2nd false.**

$(\exists x)Mx \to Hm$ ✓
$\sim[(\exists x)Mx \to (x)Hx]$ ✓
$(\exists x)Mx$ ✓
$\sim(x)Hx$ ✓
$(\exists x)\sim Hx$ ✓

```
        ~(∃x)Mx             Hm
           x                ~Ha
                            Mb
```

**2nd does not ∴ 1st: tree of the 2nd and denial of the 1st yields open branch; NOT EQUIVALENT.**

5. $\sim(\sim a=b \lor b=a)$; $Mba$

$\sim(\sim a=b \lor b=a)$ ✓
$\sim Mba$
$\sim\sim a=b$
$\sim b=a$
$a=b$      *
$\sim b=b$
x

**1st ∴ 2nd: tree of 1st and denial of 2nd yields all closed branches—showing it's impossible for the 1st to be true and the 2nd false.**

**Mba**
~~(~a=b v b=a) ✓
　~a=b v b=a ✓

　~a=b　　　b=a *
　　　　　　Maa

**2nd does not ∴ 1st: tree of 2nd and denial of 1st results in an open branch; NOT EQUIVALENT.**

7. ~Ms & (x)(~x=s → Mx); ~[(∃x)(~x=s & ~Mx) v Ms]

~Ms & (x)(~x=s → Mx)　✓
~~[(∃x)(~x=s & ~Mx) v Ms] ✓
(∃x)(~x=s & ~Mx) v Ms　✓
~Ms
(x)(~x=s → Mx)　　　　*

(∃x)(~x=s & ~Mx) ✓　　Ms
~a=s & ~Ma ✓　　　　　x
~a=s
~Ma
~a=s → Ma ✓

~~a=s　　　Ma
x　　　　　x

**1st ∴ 2nd: tree of 1st and denial of 2nd results in all closed branches—showing it's impossible for the 1st to be true and the 2nd false.**

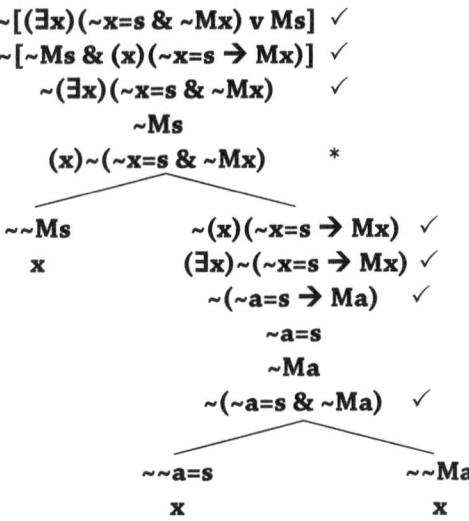

**2nd ∴ 1st:** tree of 2nd and denial of 1st results in all closed branches. Mutual Implication: **EQUIVALENT.**

# UNIT THREE: IS IT VALID?

# Patterns and Properties of Arguments

# Chapter Eleven: Arguments, Trees, and Tables

> **What's Up?**
> What's an argument?
> Validity
> Common Argument Patterns
> Short Form Tables

In previous chapters, we've looked at single statements and an important property they have: truth. We've looked at statement pairs and some relations that can hold between them: if they are consistent, contrary, contradictory, or if one implies the other, or they are equivalent. In this and the next two chapters, we'll be looking at important statement *sets* called *arguments*.

## DEFINING AND SYMBOLIZING AN ARGUMENT

An argument is defined as *a set of statements, some of which, called premises, are put forth in support of a remaining statement, called the conclusion*. Arguments give us rational or logical justification for accepting some statement (the conclusion) as true. Distinguishing arguments from non-arguments is much harder than it might first seem. This is further complicated by there being different *kinds* of arguments, and different ways to evaluate each kind. Informal Logic or Critical Thinking courses cover these issues, but in this text, using **LOLA**, we are only concerned with a particular kind of argument, called a *deductive argument*. In order to understand what we mean by a deductive argument we will present three arguments using Example Set 11a, along with the concepts of consistency and implication.

---

**Example Set 11a**
TRANSLATION SCHEME
RD: Everything

| b: Bob | d: Bob's dog | Fx: x has fleas | Px: x is a person |
|---|---|---|---|
| Cx: x brought fleas with them on their clothes. | Rx: We should check x for rabies | Ix: x is itching | Bxy: x bit y |

ARGUMENT #1: If Bob's dog bit someone, then he should be checked for rabies. Bob's dog did bite someone. Hence, we should check Bob's dog for rabies.

ARGUMENT #2: Either Bob's dog has fleas or someone brought them in on their clothes. No one brought them. So, Bob's dog has to have fleas.

ARGUMENT #3: If Bob's dog has fleas, then Bob would be itching. Bob is itching. We can conclude that Bob's dog has fleas.

The first step is to translate the arguments into LOLA. In each argument we have three statements, two of which are the premises and one of which is the conclusion. Don't worry about which is which right at the moment, simply take Argument #1 and begin translating the statements.

The first is a conditional statement:

$$(\exists x)(Px \;\&\; Bdx) \rightarrow Rd$$

The next asserts that the antecedent of the conditional is true:

$$(\exists x)(Px \;\&\; Bdx)$$

The last statement asserts the consequent: that we should check the dog for rabies:

$$Rd$$

We now need to ascertain which of these statements is the conclusion and which ones are the premises. We won't go into all the intricacies of how arguments occur in natural language; for our purposes in this text all you really need to be able to do is distinguish the premises from the conclusion. In the Unit 3 review, you'll find a list of common premise and conclusion "indicators"—those words that are often used to indicate what in an argument is the premise and what the conclusion. Not all English arguments use such words, but many do, and it's a good place to start. If you consult that list, you'll see that the word "hence" is a conclusion indicator term, and so the third statement is the conclusion of this argument. This means that the two remaining statements are the premises. But you can tell that this is an argument without consulting the list. It's clear that the speaker here intends to support the third sentence by means of asserting the first two.

There are two different ways we'll represent an argument once it's translated into LOLA. The first is called the *Horizontal Version* because it places the entire argument in a single line. Just write down the first premise, put down a comma, write down the next premise, put down a comma, and so on until you reach the conclusion, which you put inside a box. This gives us:

$$(\exists x)(Px \ \& \ Bdx) \rightarrow Rd, (\exists x)(Px \ \& \ Bdx), \boxed{Rd}$$

We'll return to this argument later, but for now we'll translate the remaining two arguments.

Argument #2 begins with a disjunction. Recall that it's easiest when translating disjunctions to simply pick one of the disjuncts, translate it, turn to the other disjunct, translate it, and then place a wedge in between them. The left disjunct says, "Bob's dog has fleas" which would be symbolized as Fd. The other disjunct is, "Someone brought them (the fleas) into the house with them." This would translate as

$$(\exists x)(Px \ \& \ Cx)$$

Putting the two together with the wedge yields,

$$Fd \ v \ (\exists x)(Px \ \& \ Cx).$$

The next statement says that nobody brought fleas in with them, and so we have the following:

$$\sim(\exists x)(Px \ \& \ Cx)$$

The last statement states that Bob's dog has fleas and that would be symbolized as

$$Fd$$

Since the final statement is introduced by the word "So" and since that is listed as a conclusion indicator term, we can see that it is the conclusion. When we set up the entire argument it looks like this,

$$Fd \ v \ (\exists x)(Px \ \& \ Cx), \sim(\exists x)(Px \ \& \ Cx), \boxed{Fd}$$

We'll just present the last argument—but be sure and check to make sure you understand why we symbolized it as we did.

$$Fd \rightarrow Ib, Ib, \boxed{Fd}$$

## DEDUCTIVE ARGUMENTS, VALIDITY, AND INVALIDITY

As we said in the beginning of the chapter, we will be focusing in this chapter on what are called *deductive arguments*. As we just explained, an argument is divided up into premises and a conclusion, with the premises being those statements put forth in support of the conclusion. Deductive arguments are ones that purport to have the following special feature: *if the premises were true then the conclusion must be true as well*. Another way to capture this is to say, *it is impossible for the premises to be true and the conclusion to be false*. Or another way, *the conclusion of a deductive argument will be true if the premises are true*. If the argument has this feature then it's said to be *valid*. If it doesn't, then it's said to be *invalid*.

> An argument is valid if the premises imply the conclusion.

Two points need to be made here. First, validity *is not a claim about the actual truth-values of the statements* that make up the premises or the conclusion—in other words, it's not a claim about the *properties* of the individual statements. Rather, *validity is a claim about the nature of the relationship among the statements that make up an argument*. In this way it is a continuation of what we began in Chapter Nine, where we began working with relationships *between* statements. As we pointed out there, to say that two statements are equivalent does not commit you to saying that the statements are true or false, it simply states that it's not possible for their truth-values to differ. So, if one statement is true then its equivalent statement *must* be true and if one statement is false then its equivalent *must* be false.

We saw something similar with the notion of implication. To say that some statement implies another statement does not commit you to saying that the statements are in fact true or that they're false; though that may be the case. Instead what implication says is that *if* the first is true then the second *must* also be true. We saw this in Chapter Nine with the table:

| A | B | ~ (~A v A) | ~A → B |
|---|---|---|---|
| T | T | F F T | F **T** |
| T | F | F F T | F **T** |
| F | T | F T T | T **T** |
| F | F | F T T | T **F** |

We saw here that ~(~A v A) ∴ ~A → B because there is no case where the first statement is true and the second is false. Since a truth table gives us all possible combinations of truth values, we know it is impossible for the first to be true and the second false. So again the point is this: implication is *not* about the truth *property* of statements, it is about the *relationship* between the statements.

The second point is related to the first. *Validity is a property of an argument, of the whole argument. This property depends, as we said above, on the relation between the premises and the conclusion. The relation in question is just that relation of implication we saw before.* But now, instead of just looking at only two statements, we look to see if the set of premises (and there could be 1, 2, 10, or any number of premises) implies the conclusion. A valid argument is one in which, when the premises are all true, then the conclusion has got to be true as well. Or again, *a valid argument is one in which the premises imply the conclusion. Or, an argument has the property of validity if the premises imply the conclusion.* As we saw before, one statement will either imply another or it won't. The same is true of the property of validity. An argument is either valid or it's not—it has to be one or the other and it can't be both at the same time.

The practical upshot here is that *we can prove an argument is valid by showing that the premises imply the conclusion.* Moreover, we can use the same methods we've already learned to do this, with some minor adjustments.

Let's look first at trees. Remember we set up a tree for implication by listing the first statement and the negation of the second statement, and then worked through the tree to see if all branches close. This shows that it's impossible for the first statement to be true and the second statement to be false. We set up a tree to determine validity in the same way, except *we list all the premises and the negation of the conclusion*, then work the tree. We'll demonstrate on the first argument.

Our first argument was:

$$(\exists x)(Px \,\&\, Bdx) \to Rd, (\exists x)(Px \,\&\, Bdx), \boxed{Rd}$$

We can set up the tree by displaying the argument by the *Vertical Method*. This way of showing the argument is to write the premises one after the other, and then after the last premise write the conclusion, in its box, to the right of the last premise, like this:

$$(\exists x)(Px \,\&\, Bdx) \to Rd$$
$$(\exists x)(Px \,\&\, Bdx) \qquad \boxed{Rd}$$

The *conclusion is next to the tree, but not on the tree*. The box reminds us of a couple of things. First, it reminds us *this is a tree to determine the validity of an argument*.

Second, it reminds us that we don't use the conclusion in running the tree: *what's in the box stays in the box.*

The next step is to write the *negated* conclusion beneath the last premise:

$$(\exists x)(Px \ \& \ Bdx) \rightarrow Rd$$
$$(\exists x)(Px \ \& \ Bdx) \qquad \boxed{Rd}$$
$$\sim Rd$$

Now we run a truth tree as we've done all along to see if the tree remains open or if it closes. If there's an *open* branch, then we've shown that it *is* possible for the argument to have all true premises and a false conclusion, and therefore it is *invalid*. If all branches close, then it means there is *no* interpretation that makes the premises true and the conclusion false, making it a *valid* argument. The completed tree follows.

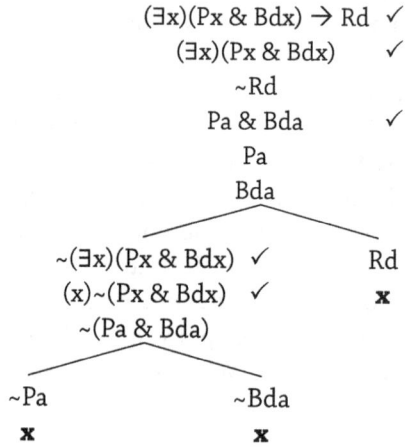

Since the tree closes it demonstrates that *there is no possible interpretation that makes the premises come out true and the conclusion come out false*. This means the argument is *valid*.

The second argument was:

$$Fd \ v \ (\exists x)(Px \ \& \ Cx), \ \sim(\exists x)(Px \ \& \ Cx), \ \boxed{Fd}$$

Set up the argument in the Vertical method, put the negated conclusion beneath the last premise, and run the tree.

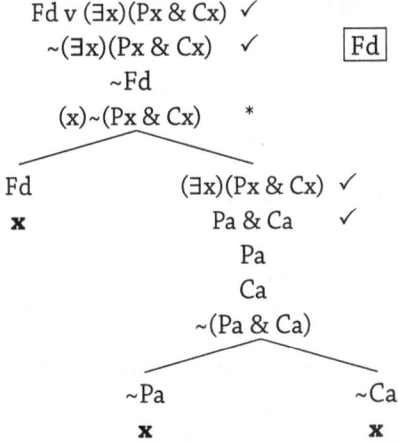

Again, since all branches of the tree close, we have established that the argument is *valid*: there is no interpretation that makes the premises come out true and the conclusion come out false.

The tree for the last argument would be as follows:

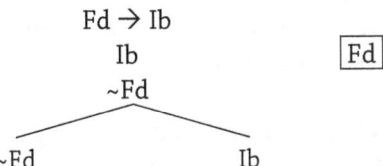

This tree, unlike the previous two trees, has open branches. This shows that it *is* possible to construct an interpretation in which the premises are true and the conclusion is false. This means that the argument is *invalid*.

## Exercise 11a

Symbolize the following arguments and then use Trees to determine if the arguments are valid or invalid.

**TRANSLATION SCHEME**
RD: People

| s: Superman | c: Clark Kent | b: Batman | r: Bruce Wayne |
|---|---|---|---|
| Hx: x is a superhero | Mx: x is mild-mannered | Px: x is a playboy | Lx: x is lying |
| Sxy: x is stronger than y | Dx: x is a great detective | Vx: x is vulnerable to kryptonite | Wx: x has a fatal weakness |

1. Batman is a superhero. Superman is a superhero. Therefore, Batman and Superman are superheroes.
2. If Clark Kent is identical to Superman, then he's a superhero. Clark Kent is the secret identity of Superman. So, Clark Kent is a superhero.
3. Superman is stronger than Batman. Therefore, either Superman is stronger than Batman or someone is lying.
4. If Batman is identical to Clark Kent, then Clark Kent is identical to Bruce Wayne. Bruce Wayne is not identical to Clark Kent. Therefore, Batman is not identical to Clark Kent.
5. Batman is a great detective, even though Superman is stronger than everyone. So, it's still appropriate to conclude that Batman is a great detective.
6. Either Bruce Wayne is a playboy or he's lying. If he's lying then he's Batman. If he is a playboy, then he's no superhero. So, either Bruce Wayne is the Batman or he's not a superhero.
7. Either Batman is a great detective or he has superhuman strength. He doesn't have superhuman strength. So, he's a great detective.
8. If Clark Kent is identical to Superman, then Clark Kent is vulnerable to kryptonite. If Clark Kent is vulnerable to kryptonite, then he has a fatal weakness. So, Clark Kent has a fatal weakness.
9. Superman is not mild-mannered, although Clark Kent is. If Superman is not mild-mannered, and Clark Kent is, then Superman is not identical with Clark Kent. Therefore, Superman and Clark Kent are not identical.
10. Clark Kent is the secret identity of Superman and Bruce Wayne is the secret identity of Batman. Superman is stronger than Batman. Hence, Clark Kent is stronger than Batman.

## ARGUMENT PATTERNS

It is an interesting and important fact of reasoning that we reason about many things—God, jobs, foreign policy, who should do the dishes—the list is endless. The *ways* we argue, however, show reoccurring patterns. A more formal way of saying this is that for any *particular pattern* of argument, there are any number of natural language *instances* of that argument. One of the advantages of studying arguments symbolically is that you can more easily recognize these recurring patterns when you encounter them. You've already seen one common recurring pattern. Argument #1 from Example Set 1 and Argument #2 above in Exercise 11a are both the same pattern. Once you've symbolized an argument the pattern becomes clear.

| **Argument #1** | **Argument #2** |
|---|---|
| 1. (∃x)(Px & Bdx) → Rd, | 1. c=s → Hc, |
| 2. (∃x)(Px & Bdx)    \|Rd\| | 2. c=s    \|Hc\| |

Now content aside (one argument is about biting rabid dogs and the other is about superheroes), there are clearly still *formal* differences between the two arguments. The first one includes existential quantifiers and conjunctions, while the second one has neither, but it does have identity statements. Here the ability to *see* main operators is very helpful in sorting things out. *The pattern of any argument is determined by the kinds of statements that make it up. For statements with operators, the kind of statements they are is determined by their main operators.* If the main operator is (x), it's a universal quantification; if it's a &, it's a conjunction, etc.

When you look at the two examples above, you can see that the main operator in the first premise for both of them is the →; it's a conditional statement. It might seem as though the second premise in both arguments is different. In one way it is, of course. They have different main operators; one is an existential quantification, the other is an identity statement. But look closer—premise 2 in both cases is a statement that is *identical* to the *antecedent* of the conditional in premise one. Finally, the conclusion for both arguments is a statement that is *identical* to the *consequent* of the conditional in premise one.

This particular pattern in which

- one premise is a conditional,
- one premise is identical to the antecedent of that conditional, and
- the conclusion is identical to the consequent of the conditional

is an extremely common argument pattern. It is so common in fact that it even has a name, *Modus Ponens*. We can represent this pattern symbolically as follows:

**Modus Ponens (MP)**

$X \rightarrow Y$
$X \quad \boxed{Y}$

The Xs and Ys just stand for statements of any complexity. This argument pattern shows only the main connective of the argument. The *Modus Ponens* pattern *always* consists of three statements: two premises and a conclusion. In one premise, the X stands for the antecedent of the conditional, whether it is a single statement or a complicated statement with two or many components. The antecedent then reoccurs as the other premise—these two occurrences must be identical. In the first premise, the Y stands for the consequent of the conditional; again, it could be a simple statement or a complex statement. It also reoccurs as the conclusion.

> **PRACTICE**
>
> As we said, this is a very common argument pattern—it has countless English languages instances. We've just given you two, about dogs and Superman. Take a few minutes and see if you can construct a few more examples that use this same pattern.

We said above that studying these patterns is useful because it helps you more quickly and easily recognize these patterns when you come across them in life. But so what? The point is this: the *Modus Ponens* pattern is *valid*. We showed the "rabid dog" instance was valid; you can run this basic "$X\,Y$" pattern on a tree and verify for yourself that it's valid. In fact *any instance of Modus Ponens is valid*. So, when you encounter an instance of *Modus Ponens* in your studies, you know that that argument instance is valid. *Remember*: you'd still need to check if the premises of the argument are *in fact true*, but at least you'd already know that *if they are in fact true, then the conclusion must also be true*.

For the remainder of this chapter we will introduce seven more common argument patterns, and then show another way of determining validity using truth tables. *Each of the following patterns is a valid argument pattern*. We will present three different versions of each one so that you can see the way it occurs in various guises. Use the following translation scheme for the first four argument patterns and we will then provide a new translation scheme for the remaining three.

Rather than begin with the representation of the argument form as we did with *Modus Ponens*, we'll give you three arguments first, and then reveal the pattern afterwards. In order to get some practice, don't skip right to the representation of the argument form. Try translating the arguments yourself first, and then identify which statements are premises and which is the conclusion. Finally, look closely and see if you can recognize the underlying pattern before checking the representation to see if you got it right.

**TRANSLATION SCHEME**
RD: Everything

| j: Jane | r: Robert | s: Sam | f: '67 Ford Mustang |
|---|---|---|---|
| m: Robert's motorcycle | w: Washington, DC | n: New York City | c: Chicago |
| Px: x is a person | Fx: x is fast | Lxy: x is larger than y | Qxy: x is quicker than y |
| Rxyz: x takes y to z | | | |

# MODUS TOLLENS (MT)

ARGUMENT #1: If Robert took the '67 Ford Mustang to New York, then he'd be quicker than Sam. He wasn't quicker than Sam. So, we can conclude that he didn't take the Ford.

ARGUMENT #2: It's not true that Chicago is larger than every city. If Chicago was bigger than New York, then it would be bigger than every city. So, Chicago isn't bigger than New York.

ARGUMENT #3: We should conclude that it's not true that there is someone who is faster than a '67 Ford Mustang. Here's why. If someone is faster than a '67 Ford Mustang, then someone is faster than Robert's motorcycle. However, no one is faster than Robert's motorcycle.

You'll find that the pattern for these arguments is as follows:

$$\text{Modus Tollens}$$
$$X \rightarrow Y$$
$$\sim Y \qquad \boxed{\sim X}$$

You may have noticed in these examples that the premises can come in any order, the author may use synonyms (e.g., bigger and larger), and the conclusion can even come first. These are merely stylistic variations; from a logical point of view, they are superficial differences. The important logical pattern is that:

1. There are two premises
2. One premise is a conditional.
3. One premise is a denial of the consequent of the conditional.
4. The conclusion is a denial of the antecedent of the conditional.

## CONJUNCTION (CONJ)

ARGUMENT #1: Someone is quicker than Robert. Furthermore, Sam is larger than Robert. We can thereby see that someone's quicker than Robert and Sam's larger than Robert.

ARGUMENT #2: Nobody is quicker than everyone. Some people are fast. This proves that even though nobody is quicker than everyone, there are some fast people.

ARGUMENT #3: Chicago is larger than Washington, DC. New York is larger than Washington, DC. All this shows that both New York and Chicago are larger than Washington, DC.

The pattern for this argument is:

$$\begin{array}{ccc} & \text{Conjunction} & \\ X & & Y \\ Y \quad \boxed{X \& Y} & \textbf{OR} & X \quad \boxed{X \& Y} \end{array}$$

Either one is an acceptable version of Conjunction.

## SIMPLIFICATION (SIMP)

ARGUMENT #1: Sam is quicker than Robert, though he's not quicker than Jane. This at least shows that Sam's quicker than Robert.

ARGUMENT #2: Both New York and Chicago are fast. This allows us to infer that Chicago is fast.

ARGUMENT #3: Jane took the '67 Ford Mustang to New York, but she took Robert's motorcycle to Chicago. Therefore, Jane took the bike to Chicago.

Notice that these arguments have only *one* premise and a conclusion. This is not a problem, not all argument patterns have to have a two premise, one conclusion structure.

This pattern is:

$$\begin{array}{ccc} & \text{Simplification} & \\ X \& Y \quad \boxed{X} & \textbf{OR} & X \& Y \quad \boxed{Y} \end{array}$$

Again, either one is an acceptable version of Simplification.

## ADDITION (ADD)

ARGUMENT #1: Jane is fast. As a result you can see that either Jane is fast or Robert's fast.

ARGUMENT #2: Nothing is larger than New York. Hence, nothing is larger than New York or everything is larger than New York.

ARGUMENT #3: Robert's bike is faster than everything. Therefore, Robert's bike is quicker than everything unless Jane's Ford is faster than Robert's bike.

This pattern is:

$$\text{Addition}$$
$$X \quad \boxed{X \lor Y} \quad \textbf{OR} \quad X \quad \boxed{Y \lor X}$$

Either one is an acceptable version of Addition.

For the last three argument patterns provide a different translation scheme. Again, try to determine the pattern yourself before going directly to our representation.

### TRANSLATION SCHEME
RD: Countries

| u: United States | r: Russia | c: Canada | h: Haiti |
|---|---|---|---|
| e: Eastern Europe | Gx: x has a large gross national product | Mx: x spends a great deal on military budget | Hx: x spends a great deal on health care |
| Cx: x has been racked with civil conflict | Fx: x improves its foreign relations | Rx: x must reform its internal political system | Dxy: x dominates y |
| Axyz: x is a closer ally than y is to z | Nxyz: x is nearer to y than z | | |

## DISJUNCTIVE SYLLOGISM (DS)

ARGUMENT #1: Either the United States is closer to Russia than Canada or Haiti is closer to Russia than Canada. Haiti is not closer to Russia than Canada. This proves that the United States is closer to Russia than Canada.

ARGUMENT #2: Either Russia spent a great deal on its military budget or it would be racked with civil conflict. It hasn't been racked with civil conflict. We can conclude that it spent a great deal on its military budget.

ARGUMENT #3: Obviously, Canada has a large gross national product. Either Canada has a large gross national product or there's no country with a large gross national product. But of course, it's not the case that no country has a large gross national product.

This argument pattern is:

Disjunctive Syllogism

$X \vee Y$　　　　　　　　　　$X \vee Y$
$\sim X$　$\boxed{Y}$　　**OR**　　$\sim Y$　$\boxed{X}$

## DILEMMA (DI)

ARGUMENT #1: Either Russia is going to dominate Eastern Europe or the US will. If Russia is to dominate it, then it will need to reform its internal political system. If the US is to dominate it, then it will need to improve its foreign relations. Consequently, either Russia will reform its internal political system or the US will improve its foreign relations.

ARGUMENT #2: Either Canada will spend a lot on medicine or it won't. If it does then it can't spend a great deal on its military budget. If it doesn't then it will not be able to improve its internal political system. Hence, it either won't improve its internal political system or it won't spend much on its military budget.

ARGUMENT #3: Haiti will improve its foreign relations unless it suffers from a civil conflict. If it suffers from civil conflict then it will need to spend a lot on its military budget. If it improves its foreign relations then it will be a closer ally to the US than Russia. As a result, either Haiti will spend a great deal on its military budget or it will become a closer ally to the US than Russia.

Notice that this pattern has *three* premises and a conclusion:

Dilemma
$X \vee Y$
$X \rightarrow W$
$Y \rightarrow Z$　$\boxed{W \vee Z}$

## HYPOTHETICAL SYLLOGISM (HS)

ARGUMENT #1: If the US dominates Russia, then it has to spend a great deal on its military budget. If the US spends a great deal on the military budget, then it can't have a large gross national product. So, if the US dominates Russia, then it can't have a large gross national product.

ARGUMENT #2: If Russia is racked with civil conflict, then there is at least one country that has a civil conflict. If there is at least one country that has a civil conflict, then there is at least one country that has to spend a lot on its military budget. Therefore, if Russia is racked with civil conflict, then there is at least one country that has to spend a lot on its military budget.

ARGUMENT #3: If the US is nearer to Russia than it is to Canada, then it will be a closer ally to Russia, than it will be to Canada. If it's a closer ally to Russia than it is to Canada, then it will dominate all countries. Hence, if the US is nearer to Russia than it is to Canada, then it will dominate all countries.

This pattern takes the following form:

$$\text{Hypothetical Syllogism}$$
$$X \rightarrow Y$$
$$Y \rightarrow Z \quad \boxed{X \rightarrow Z}$$

## Exercise 11b

Symbolize the arguments below, give the name of the argument form, and then construct a tree to demonstrate that the argument is valid. All the argument forms are represented at least once and a couple of the argument forms show up more than once.

**TRANSLATION SCHEME**
RD: Everything

| b: Bob | j: John | Px: x is a person |
|---|---|---|
| Fx: x is free | Dx: x is dull | Ux: x is fun |
| Cx: x is confused | Bx: x will be bored | Fxy: x wants to be friends with y |
| Ix: x needs to fight | | |

1. If at least one thing is fun, then there's a least one thing that isn't dull. If there is at least one thing that isn't dull, then a person doesn't have to be bored. Hence, if at least one thing is fun, then people don't have to be bored.
2. If John is free, then he needs to fight. John is free. Accordingly, he needs to fight.
3. John is not fun. If he wants to be friends with Bob, then he should be fun. This leads me to believe that John doesn't want to be friends with Bob.
4. Either everyone is fun, or someone is dull. It's not the case that everyone is fun. Consequently, someone is dull.
5. It's definitely the case that nothing is free. You must conclude that either Bob is hopelessly confused or nothing is free.
6. John wants to be friends with everyone, but Bob wants to fight. You still have to conclude that John wants to be friends with everyone.
7. Everyone will be bored. Everyone is also going to fight. Unfortunately, this implies that everyone will be bored and everyone is going to be fighting.
8. Either everyone is free or no one is free. If everyone is free, then we don't need to fight. If no one is free, then everyone is confused. Thus, either we don't need to fight or everyone is confused.
9. Someone wants to be friends with John, which we can see for the following reasons. If someone is fun, then they'll want to be friends with John. There has to be someone out there who is fun.
10. Some people want to fight or everyone is bored. Not everyone is bored. It follows that some people want to fight.

## VALIDITY AND SHORT-FORM TABLES

We can also use truth tables to determine whether an argument is valid or not, although truth tables have limitations. As we saw in Chapter Seven, truth tables are not able to handle predications, quantifications, or identity statements.

But as we also saw, we *can* use tables for those statements that have one of the five connectives as their main operators, and that we can adequately capture in LOLA-Lite. Once we represent the English language statements in this fashion, we can then use truth tables to determine properties of statements (L-true, L-false, contingency) or the relation between statements (consistency, contrariety, contradiction, implication, equivalence).

We can extend this process to arguments as well, although they must be composed of statements that can be adequately represented by symbolizing their component statements as either single capital letters, or capital letters with connectives. Of course, many arguments need a full translation of the quantifications and predicate statements in order to show their validity. The short arguments presented in this chapter (instances of MP, MT, DS, etc.) are just the kinds of arguments that a truth table can handle very well. This is why being able to move back and forth between LOLA and LOLA-Lite can be handy.

But (you may be thinking) why a truth table, when we have trees? Truth tables, as we saw, get unwieldy if you have more than 3 sentence letters, and maybe most arguments would have at least that many. Good point. But to determine validity, we don't need to do a full truth table. We only have to find a row, *if it exists*, where the premises are all true and the conclusion is false—if we found such a row we'd know the argument was *invalid*. If such a row does *not* exist, then we'd know the argument was *valid*.

But (you may be thinking) how do we know in advance where to find this row without doing the complete truth table? Another excellent question. And the answer is: we use the technique of *Short-Form Tables*.

Short-form tables are fairly easy to set up and to work, but it helps to review what we know about truth tables and the concept of implication to start. Remember, that to use a table to check for implication we start at the top row by writing down the individual statement letters, then the first statement, then draw a double line, and write down the second statement. You then fill in all the values for the atomic and compound statements, complete the truth table, and check to see if the first statement implied the second. You do this by seeing if there is any row in which the first statement reads true, and the second statement reads false. If there is even one row, then we say that the first statement does *not* imply the second statement. If there is no such row, then the first statement *does* imply the second.

Given the similarity between the concepts of implication and validity, it's not surprising how similar they are when it comes to truth tables. Validity can be defined as a situation in which the premises imply the conclusion. If an argument has only one premise and one conclusion (e.g., SIMP or ADD), a truth table to show validity would look exactly the same as a truth table to determine implication. In a truth table for an argument with more than one premise, the only difference is that *each* premise is written on the first row before the double line, and the conclusion is written after the double line. We'll illustrate the use of truth tables and short-form tables using the following examples:

**Example Set 11b**
1. If olives are healthy, then I can use olive oil on my salad. If I use olive oil on my salad, then I won't feel deprived. So, if olives are healthy, I won't feel deprived. (O: Olives are healthy; I: I use olive oil on my salad; D: I feel deprived.)
2. I can go out to eat and go to a movie unless I don't get paid. But I will get paid. So, it's dinner and a movie for me! (E: I go to eat; M: I go to the movie; P: I get paid.)
3. If everyone accepts the economic proposal, then some people will suffer. And everyone does accept it. So, I guess some folks are going to suffer. (A: Everyone accepts the economic proposal; S: Some people will suffer.)
4. Voldemort won't be defeated! Voldemort would be defeated if both Harry and Fudge fought him. But you know that Harry and Fudge will never get together to fight him. (V: Voldemort will be defeated; H: Harry fights Voldemort; F: Fudge fights Voldemort.)

Before we try a short-form table, we'll test the first two arguments for validity by using complete truth tables. Begin as always by symbolizing the argument:

**Argument 1: O → I, I → ~D** $\boxed{O → ~D}$

Now set up the truth table. There are 3 sentence letters, so the table will have 8 rows. Be sure to put the conclusion at the end of row one, and a double line before it.

| O | I | D | O → I | I → ~D | O → ~D |
|---|---|---|---|---|---|
| T | T | T | T | F | F |
| T | T | F | T | T | T |
| T | F | T | F | T | F |
| T | F | F | F | T | T |
| F | T | T | T | F | T |
| F | T | F | T | T | T |
| F | F | T | T | T | T |
| F | F | F | T | T | T |

As with implication, the only rows we're concerned with are those where the premises, *all the premises*, are true. In this table these are rows 2, 6, 7, and 8.

In a valid argument, if the premises are true, it's *impossible for the conclusion to be false*. That's what we see here: in no case, when all premises are true, is the conclusion false.

**VALID ARGUMENT**

*Argument Form: Hypothetical Syllogism*

## Argument 2: (E & M) v ~P, P ∴ E & M

| E | M | P | (E & M) v ~P | P | E & M |
|---|---|---|---|---|---|
| T | T | T | T | T | T |
| T | T | F | T | F | T |
| T | F | T | F | T | F |
| T | F | F | T | F | F |
| F | T | T | F | T | F |
| F | T | F | T | F | F |
| F | F | T | F | T | F |
| F | F | F | T | F | F |

Here, there's only one row where both premises are true; so, this is the only row that concerns us.

The conclusion is also T. This mean that the premises, taken together, imply the conclusion, OR: The argument is **VALID**.

Now we'll demonstrate the short-form table technique using Arguments 3 and 4. One way to think about this technique is that, in short-form, we do what we do in full tables, but we do it backwards! Instead of "discovering" rows that prove or disprove validity, we set about trying to *create* a row where the conclusion is FALSE, and all the premises are TRUE.

We begin to do this by assigning truth values—but not to the individual capital letters as in the full table. Instead we *assign truth values to the conclusion that makes it false*. Then we fill in those same values where they occur in the premises, and make other assignments as needed until we've made an assignment to each letter in the premises. If we *can* make assignments that make the premises come out true, while the conclusion is false, we know the argument is *invalid*. If it is *impossible* to make the premises come out true, while the conclusion is false, we know the argument is *valid*. This will become clearer with an example.

## Argument 3: A → S, A ∴ S

You should recognize that this is an instance of *Modus Ponens*. So, we already know this is valid, but we can prove it again, this time with a short-form table.

First, set up the first row as you normally would.

| A | S | A → S | A | S |
|---|---|---|---|---|

Now, we assign truth-values to make the conclusion FALSE. In this case, this is easy. There's only one way to make the conclusion false—by assigning the value F to the letter S. We'll make another row under the first, and put an F in the conclusion column, as well as in the statement letter column under S, and where S appears in the premises.

| A | S | A → S | A | S |
|---|---|---|---|---|
|   | F | F |   | F |

We need now to determine if it is possible to make both premises true. Usually, if one of the premises is a single statement letter, start there. You can make the second premise true just by assigning the value T to the letter A. We'll fill this in, in the first column as well as under the A in the premise A → S.

| A | S | A → S | A | S |
|---|---|---|---|---|
| T | F | T  F | T | F |

But now look at premise 1. We have to assign A the value of true in order to make the second premise true. We've already assigned S the value of false, when we were working on the conclusion. However, the result is that the first premise, A → S, is false; any conditional where the antecedent is true and the consequent false means the conditional as a whole is false. (If we started with assigning values to make the first premise true, then the second premise would be false—try it!)

| A | S | A → S | A | S |
|---|---|---|---|---|
| T | F | T F F | T | F |

What we have shown is that it is *impossible for both premises to be true and the conclusion to be false*. Therefore, the argument is **VALID**.

### Argument 4: (H & F) → V, ~(H & F)  ~V

Again, set up the first row, and then assign truth values to make the conclusion false. In this case, we need to assign V the value T, so that ~V will be false.

| H | F | V | (H & F) → V | ~(H & F) | ~V |
|---|---|---|---|---|---|
|   |   | T | T |   | F T |

Now we have some choices. Premise #1 is a conditional with a true consequent. This means that the conditional as a whole is true, no matter what values we assign to H and F. So, let's move onto the second premise which is a *negated* conjunction. We want the premise to be true, which means the conjunction needs to be false. Conjunctions are false when either H or F or both of them are false. We'll try making both of them false, which makes the conjunction false, and which, when it's negated, makes premise #2 true.

| H | F | V | (H & F) → V | ~(H & F) | ~V |
|---|---|---|---|---|---|
| F | F | T | F **F** T T | **T** F F F | **F** T |

And we're done. We've shown we can create a row, by assigning truth values, where the premises are all true, but the conclusion is false. So this argument is **invalid**.

As a matter of fact, there are other assignments of values that will result in true premises and a false conclusion in the complete truth table. But remember—*validity is an all or nothing property*. All we need to show is *just one instance* when true premises lead to a false conclusion, and we know the argument is invalid. We'll show you one more row, though, just for more practice.

| H | F | V | (H & F) → V | ~(H & F) | ~V |
|---|---|---|---|---|---|
| F | F | T | **T** | **T** | **F** |
| F | T | T | F **F** T T | T F **F** T | **F** |

Again, all true premises and a false conclusion. This argument is, again, INVALID.

Before leaving this, we want to show you one complication. In some arguments there are more ways than one to make the conclusion false. In these cases, you must try each way until you definitely determine if the argument is valid or invalid. You might prove it invalid on the first try—if so, stop; your work is done. If however, you find it impossible to have all true premises and a false conclusion on one try, then you need to try other possibilities until you conclusively prove it one way or the other. We'll illustrate this with one easy example.

| A | B | A v B | A & B |
|---|---|---|---|
| F | F |   | F **F** F |
| F | T |   | F **F** T |
| T | F |   | T **F** F |

3 assignments that would make the conclusion false. We'll try the first option first.

| A | B | A v B | A & B |
|---|---|-------|-------|
| F | F | F F F | F F F |

In the 1st row with these truth value assignments, you can't make the premise true and the conclusion false. But before we declare it valid, we need to test row 2.

| A | B | A v B | A & B |
|---|---|-------|-------|
| F | F | F F F | F F F |
| F | T | F T T | F F T |

In the 2nd row with these truth value assignments, we see it IS possible to have a true premise with a false conclusion. So, this argument pattern is INVALID.

## Exercise 11c

For the following argument patterns, use short-form tables to determine if the patterns are valid or invalid. Be sure to state if each argument is valid or invalid.

1. A v B, C v D           | A v D
2. A → (B & C), ~A        | ~B v ~C
3. A → (B v ~D), A → C    | B → C
4. A ↔ B                  | ~B → A
5. A v B, A → C, B → D    | C v D
6. ~A → B                 | A & B
7. B → (A & B), (C v A) & B | A ↔ B
8. C ↔ B, ~K → ~C, ~C → ~K | B v K
9. C v B, A & ~B          | C
10. B & (C v A), B v A, A → D | D v B
11. D v (B → ~C), ~D & C  | ~B
12. ~B → (C & D), ~D      | B
13. K ↔ M, M → L, L v ~M  | M v L
14. ~C → Q, K v C         | ~K → C
15. J v B, G & ~O, J → O  | B & G
16. P & ~Q, Q v (P → M)   | M v Q
17. D ↔ C, C v K          | D v K
18. G v (H & C)           | (G & H) v (G & C)
19. N v P, P → ~N         | N ↔ P
20. A → (M & N), ~M v ~N  | ~A

# Chapter Twelve: Method of Proof

> **What's Up?**
> Constructing Proofs
> Basic Proof Rules
> Equivalence Rules
> CP and RAA

Thus far you have learned how to use truth tables and truth trees to ascertain the truth-value of statements, as well as to determine a wide variety of relationships among statements. In the previous chapter we introduced the concept of an argument and explained how to use truth trees to determine if an argument is valid or invalid, as well as how to use truth tables for LOLA-Lite arguments. As we pointed out, there are severe restrictions on what truth tables can show with regard to arguments when those statements contained universal or existential statements. However, this does not mean that the only way to demonstrate those arguments is with truth trees.

For the last section of the text we introduce you to one final technique, the method of proof. A logical proof demonstrates to everyone that a valid argument is indeed valid. This technique is able to handle any valid argument, even those with universal, existential, and identity statements. It does have one drawback though; we can't use it to determine if an argument is *invalid*, we can only show that one *is* valid. While this is certainly a disadvantage it has other features that make it well worth studying.

Now truth tables are good examples of what we can call "mechanical" thinking. We have a set of functions, we put in given inputs (Ts and Fs), and we generate outputs. If we follow the rules, we will always get the right answer. Trees rely on the same truth-functional definitions, and we proceed by asking the same question at each step ("Under what condition(s) is this statement true?"). But in trees we have choices: branch or not? Instantiate to 'a' or to 'b'? Instantiate the existential or do quantifier negation on the ~(∃x) first. There is, in other words, some ingenuity involved in finishing a tree.

Many people, however, believe that *proofs* are a better representation of the way we usually reason "out in the world." We begin with a pool of information (set of statements) and then work out the implications of this information in a step-by-step manner (though sometimes it's a two-steps forward and one-step back manner). We see how information goes together, we detect patterns, we find ingenious ways to short-cut to our goals. This type of reasoning is much less mechanical than either trees or tables; it allows far more creativity. For complex arguments, it sometimes takes a lot of ingenuity to show validity—and we're not always successful. A further point is that proofs may be the clearest visual representation of a valid argument; it's the easiest way to **see** validity.

## CONSTRUCTING A PROOF

We will introduce the method of proof by considering three arguments using the following example set.

### Example Set 12a
**TRANSLATION SCHEME**
RD: Animals

| k: King Kong | l: Lassie | r: rover | m: Morris the cat |
| Ax: x is an ape | Dx: x is a dog | Cx: x is a cat | Hx: x is a threat to humanity |
| Sxy: x is smarter than y | Txy: x is tougher than y | Fxy: x is faster than y | |

ARGUMENT # 1: If King Kong is an ape, then he's smarter than Lassie. King Kong is an ape, but he's a threat to humanity. So, King Kong is smarter than Lassie.

The first step is to symbolize the argument in one of two ways: the Horizontal Version or the Vertical Version that we introduced in Chapter Ten.

$$\text{Horizontal Version}$$
$$Ak \rightarrow Skl, \; Ak \,\&\, Hk \quad \boxed{Skl}$$

$$\text{Vertical Version}$$
$$Ak \rightarrow Skl$$
$$Ak \,\&\, Hk \quad \boxed{Skl}$$

*We'll often give you an argument in the Horizontal Version and your first step in constructing a proof is to turn it into the Vertical Version.*

As we said earlier, the goal is to demonstrate that this is a valid argument, to show that the truth of the premises *guarantees* the truth of the conclusion. The method of proof does this by using a set of rules that sanctions the creation of new lines in the proof that are intermediate conclusions validly justified using statements occurring earlier in the proof. Any use of these rules on a premise will guarantee that the result is true, if the premise is true. Any use of these rules on something we've shown is true, given the truth of the premises, will itself be true. In other words, these rules are *truth-preserving*. This line creation continues until you can validly conclude the "boxed" conclusion. The proof is finished when the last line in the proof is the same as the boxed conclusion, because we've shown that the boxed conclusion must be true, if the premises are.

The first rules we'll present are the valid argument forms that we introduced in Chapter Eleven. Each one of those argument patterns will become a rule in our proof system. As long as we use only valid argument forms as our rules then we'll be safe; by always moving from one valid argument to another and then to another, we can't go wrong. This means that if we are able to derive the conclusion from our premises at some point down the line, we will have demonstrated that the conclusion is *implied* by the premises. Of course, this is one of the ways that we define a valid argument.

> Proofs are a series of mini-arguments, each of which is valid, which lead to a final conclusion.

Here are the first eight rules:

**Modus Ponens (MP)**
$X \to Y$
$X \quad \boxed{Y}$

**Modus Tollens (MT)**
$X \to Y$
$\sim Y \quad \boxed{\sim X}$

**Simplification (SIMP)**
$X \& Y \ \boxed{Y}$ **OR** $X \& Y \ \boxed{X}$

**Conjunction (CONJ)**
$X \qquad$ **OR** $\quad X$
$Y \ \boxed{X \& Y} \qquad Y \ \boxed{Y \& X}$

**Addition (ADD)**
$X \ \boxed{X \vee Y}$

**Disjunctive Syllogism (DS)**
$X \vee Y \quad$ **OR** $\quad X \vee Y$
$\sim X \ \boxed{Y} \qquad \sim Y \ \boxed{X}$

**Hypothetical Syllogism (HS)**
$X \to Y$
$Y \to Z \ \boxed{X \to Z}$

**Dilemma (DI)**
$X \vee Y$
$X \to Z$
$Y \to W \ \boxed{Z \vee W}$

These are called *implication rules* because the first line or lines of the rule *imply* what's in the box. However, this does not go the other way; it is *not* the case that what's in the box allows you to conclude the premises of the rule.

For the moment, these eight will be the only rules that we'll use in constructing a proof. The goal for establishing the validity of an argument is to show that, given a particular set of premises, we can derive what's in the box by using these rules. Of course, you may not use what's *in* the box in this process, since that is the very thing you're trying to derive.

Let's return to the first argument.

$$Ak \rightarrow Skl$$
$$Ak \ \& \ Hk \quad \boxed{Skl}$$

The first step is to take the vertically displayed proof and number the lines.

1. Ak → Skl
2. Ak & Hk   $\boxed{Skl}$

Remember: we're trying to get the conclusion, Skl, as the last line in our proof. Now in a proof we are allowed to use any of the rules. But what kind of rules we use depends on what kind of statements we have. Further, it helps to efficiently complete proofs by making "strategic moves." At the end of this chapter, we'll list a series of *Strategy Tips*. These are suggestions for you as to how to proceed through a proof. These aren't requirements, though; it may turn out that you have a different approach to solving a proof, and that's fine.

By checking the conclusion, Skl, and then looking to see if there's anything like it in the premises, we can narrow down which rules would be best to use. Notice that the conclusion is identical to the consequent of the conditional that's on line #1, "Ak → Skl." So, then ask yourself, "do I have a rule that lets me derive the consequent of a conditional?" And of course you do: *Modus Ponens*. To use *Modus Ponens*, you need two things:

1. a conditional (which you've got on line #1)
AND
2. a statement that's identical to the antecedent of the conditional (which, at the moment, you *don't* have)

When we look over our premise lines we do see that the conjunction on line #2 contains the statement, Ak. Remember though, Ak has to be *by itself* on a line in order to be able to use the rule, *Modus Ponens*.

In this case, our mini-goal is to get Ak by itself, which can then be used as part of *Modus Ponens*, which will then generate our conclusion.

As always, it's essential to know exactly what the main connective for a statement is. Only by knowing what the main connective is, can we know what rule to use. For example, what kind of statement is it that contains Ak? It's a conjunction.

Which of our eight rules allow you to do something with conjunctions? There's only SIMP and CONJ, and once you look at these two rules you can quickly see which one you have to use (SIMP, of course).

So, we use SIMP, which allows us to derive either of the conjuncts from a conjunction. The procedure is to begin by putting down the next number.

1. Ak → Skl
2. Ak & Hk          [Skl]
3.

Then you put down the statement that you're obtaining via your rule, in this case, Ak.

1. Ak → Skl
2. Ak & Hk          [Skl]
3. Ak

Next, you put down the rule that you've used to derive that statement and the lines that you're using when employing that rule. Some rules require you to reference three lines, some two lines, and others only one line. In this case, you're using SIMP which only references one line (because it only has one premise).

1. Ak → Skl
2. Ak & Hk          [Skl]
3. Ak                2, SIMP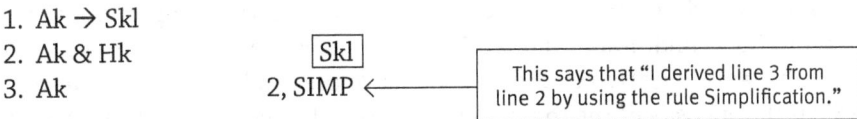

This says that "I derived line 3 from line 2 by using the rule Simplification."

Having derived the statement we were looking for, Ak, we can now use it together with the conditional on line 1 and the rule MP. Again, the procedure is to reference the lines that are used with the rule, followed by the name of the rule that is being utilized. In this case, notice that MP requires you to reference *two* distinct lines: the one with the conditional on it, and the one with statement identical to the antecedent of the conditional.

1. Ak → Skl
2. Ak & Hk          [Skl]
3. Ak                2, SIMP
4. Skl               1, 3 MP

Since you have a statement on line #4 that is identical to what's in the box, the proof is completed. We have derived a statement that is identical to conclusion, thereby showing that the premises imply the conclusion; that is, we know if the premises were true, then the conclusion would also be true.

ARGUMENT #2: If Lassie is faster than Morris, then she must be a cat. If Lassie is a cat then she's smarter than King Kong. It's not the case that Lassie is smarter than King Kong. Hence, Lassie is not faster than Morris.

1. Flm → Cl
2. Cl → Slk
3. ~Slk

As we suggested earlier a good first step is to look at the conclusion, ~Flm, and see whether there is anything like it in the premises. While there isn't a statement *exactly* like it in the premises anywhere, there is something that *sort of* resembles it. In this case, there is a statement that is identical to the *opposite* of the conclusion as the antecedent of the conditional on line #1. So, we begin by looking at all the rules for conditionals: *Modus Ponens, Modus Tollens, Hypothetical Syllogism, Dilemma*. Of these rules, which one lets us derive the negation of its antecedent? There's only one, MT. It requires both a conditional (which we have on line #1) and the negation of the consequent of that conditional on a separate line all by itself (which we don't have at the moment).

This means our intermediate goal is to derive a statement which is identical to the negation of the consequent of the conditional on line #1; so, ~Cl is a new goal. Once we have it, we'll be able to use it with the conditional on line #1 and the rule *MT* to derive the conclusion. Interestingly, ~Cl is the opposite of the antecedent of the conditional on line #2. So, we can employ MT here as well, as long as we have the negation of the consequent of the conditional. We *do* have that. ~Skl is the opposite of the consequent of the conditional on line 2, Skl. Therefore, we have our first move.

1. Flm → Cl
2. Cl → Slk
3. ~Slk
4. ~Cl          2, 3 MT

Having obtained our intermediate goal, ~CL, via MT, we can use it along with the conditional on line #1, with MT again. This gives us the completion of our proof.

1. Flm → Cl
2. Cl → Slk
3. ~Slk     ~Flm
4. ~Cl          2, 3 MT
5. ~Flm         1, 4 MT

As we were working through this proof, some of you may have thought of a different way of constructing the proof using HS, instead of using MT twice. Such a proof would look like this:

1. Flm → Cl
2. Cl → Slk
3. ~Slk          $\boxed{\text{~Flm}}$
4. Flm → Slk     1, 2 HS
5. ~Flm          3, 4 MT

This version is just as correct as the first version we constructed. This illustrates an important aspect of the method of proof. *Any* proof is acceptable as long as you follow the rules. As long as you follow the rules it doesn't matter how long or short the proof is, or what rules you use, or in what order.

We'll conclude this section by looking at one more argument.

ARGUMENT #3: King Kong is a threat to humanity. If he's a threat to humanity then he's tougher than Morris. Of course, if he's not a threat to humanity, then at least one animal is not a threat to humanity. So, either King Kong is tougher than Morris, or at least one animal is not a threat to humanity, or King Kong is tougher than Lassie.

Translation: Hk, Hk → Tkm, ~Hk → (∃x)~Hx   $\boxed{\text{(Tkm v (∃x)~Hx) v Tkl}}$

The first step is to convert it to the Vertical Version and number the lines.

1. Hk
2. Hk → Tkm
3. ~Hk → (∃x)~Hx   $\boxed{\text{(Tkm v (∃x)~Hx) v Tkl}}$

As we did in the last argument, we'll provide two different versions of the proof. Before you look at the solutions, try it for yourself. We suggest that you try one version using Add and DI, and another version using MP and Add.

Version #1

1. Hk
2. Hk → Tkm
3. ~Hk → (∃x)~Hx     $\boxed{\text{(Tkm v (∃x)~Hx) v Tkl}}$
4. Hk v ~Hk          1, Add
5. Tkm v (∃x)~Hx     2, 3, 4 DI
6. (Tkm v (∃x)~Hx) v Tkl   5, ADD

Version #2

1. Hk
2. Hk → Tkm
3. ~Hk → (∃x)~Hx          ⟦(Tkm v (∃x)~Hx) v Tkl⟧
4. Tkm                    1, 2 MP
5. Tkm v (∃x)~Hx          4, ADD
6. (Tkm v (∃x)~Hx) v Tkl  5, ADD

Before you go on to practice constructing proofs, we'll introduce our number one strategy tip.

*Strategy Tip #1: When you're stuck, just do something. Apply any rule that you can legitimately use, even if you don't know where you're going with it.*

It may seem a bit odd, but we have found that one of the most challenging aspects of proofs for students is a paralysis that sets in as they try to think of the perfect first move. Remember, as long as you use the rules correctly, any move you make will be acceptable. It may turn out that you could have done a shorter version of the proof, but that's not a problem: it doesn't matter how long a proof is; *what matters is that you get from premises to conclusion by valid steps.*

## Exercise 12a

Construct proofs for the following arguments. As you can see, all the arguments are represented in Horizontal Version. Remember, that this means the first step will be for you to represent the argument in Vertical Version.

1. Ma, Ma → (x)(Px → Kx), (x)(Px → Kx) → (∃x)Lx, (∃x)Lx → Ta   ⟦Ta⟧
2. ~~Kb, Lb → ~Kb   ⟦~Lb⟧
3. (Nd & Pd) → Om, ~Bc → (Nd & Pd), ~Om   ⟦~~Bc⟧
4. ~Qrb → ~Ne, ~Qrb   ⟦~Ne v Ka⟧
5. Ct → (Ts → Hs), ~(Ts → Hs), ~Va v Ct   ⟦~Va⟧
6. Ba → (Gkl → Da), ~Ba → Rm, ~(Gkl → Da)   ⟦Rm⟧
7. Np → (Fo v ~Os), ~~Os & Np   ⟦Fo v ~La⟧
8. [(∃x)Bx & (y)Zy] → Qa, (∃x)Bx, Qa → (Ci & Ce), (y)Zy   ⟦(Ci & Ce)⟧
9. ~Nr v Gr, ~Nr → Et, Gr → ~Hs, (Et v ~Hs) → (Nq → Ga), ~Ga   ⟦~Nq⟧
10. (~Vc & ~Ke) → (Be → Ke), ~Vc, Ke → Vc   ⟦~Be⟧
11. ~Dq & ~~Ba, (~Dq & Rb) → ~Ar, ~Ba v Rb   ⟦~Ar⟧
12. Fa → Am, ~Ja & ~Kb, Hn → (Ga → Fa), ~Kb → (~Ja → Hn)   ⟦Ga → Am⟧

13. Rk → Db, Bt → Rk, (Bt → Db) → (Es v Fe), ~Es   |Fe|
14. ~(Rm v Sa), ~(Tcb & Va) → (Rm v Sa), ~~(Tcb & Va) → Wa   |Wa v ~Rm|
15. (Ad v Gd) → (z)Mza, (z)Mza → (Bf → Fd), Ad & Bf   |Fd|
16. (∃x)Dx → Wd, [Ga → Wd] → (Vat & ~Tcb), Ga → (∃x)Dx, Qb → Tcb   |~Qb|
17. (~Nb & Md) → (w)Qw, ~Om → Md, ~Om & ~Nb   |(w)Qw v Sk|
18. (x)Px, (x)Px → Al, Al → (Eb v Fb), ~Eb   |Fb|
19. Pa v Qa, (Qa & ~Rg) → Sb, Rg → Pa, ~Pa   |Sb|
20. (Mp v Np) → ~Sc, Tf → (Mp v Np), ~Sc   |Tf → ~(Mp v Np)|

## INTRODUCING EQUIVALENCE RULES

A method of proof needs to be able to do two things. First, if an argument is valid, the method has to be guaranteed to demonstrate that it's valid. It can't be the case that you have a valid argument, but you're missing some rules or you lack the right kinds of rules to be able to construct a proof for it. This doesn't mean that you'll necessarily be skilled enough to construct a proof for it yourself, but it has to be the case that it would be *possible* to do it with the system. If a system has this property then it's called *complete*. The second essential feature of a system is that it must guarantee you that you won't be able to construct a proof for an *invalid* argument. If has to be the case that if you follow all the rules properly you will be able to construct proofs for *only* valid arguments. If a system has this property then it's called *consistent*. A logical system has to be both complete and consistent.

The method of proof we just introduced is consistent but it's not complete, which is one reason we will be introducing more rules. However, completeness isn't everything; we'd also like it to be efficient. There are many different logical systems with different sets of rules for constructing proofs. Some have only a few rules, others have ten, and others have over twenty. The fewer rules there are, the easier it is to become accustomed to them and understand how to work with them. The downside is that in a system with only a few rules the proofs can be quite long. On the other hand the more rules a system has the shorter the proofs, although learning them all thoroughly may be hard for some people to do. The challenge is to design an efficient system, one that has enough rules to make constructing proofs relatively easy, but not have so many rules that the system is awkward to manage.

One final point, the next set of rules will be different from the first eight that we presented in an important way. The first eight were called implication rules because the premises of the rule *imply* what's in the box. However, just as we saw when we first introduced the concept of implication, the relationship only goes one way; what's in the box does not imply the first part of the rule. These next rules are called *equivalence rules* because the two parts *mutually imply* each other. This means that the two statements are interchangeable. The way we symbolize this is by placing four dots between the two statements.

As we introduce a rule we will present a diagram of it and, in some case, give you some English variants that you are by now very familiar with. We'll follow with a symbolic argument that utilizes the new rule. Sometimes we will add some comments about the rule, alerting you to potential difficulties in using the rule. After we have introduced four rules we'll have an exercise that uses the first eight implication rules and the four new equivalence rules. We will then introduce four more equivalence rules followed by another exercise that uses all sixteen rules. Use the following translation scheme for the next four rules.

**TRANSLATION SCHEME**
RD: Everything

| m: Murray | j: Janice | l: Lori | i: vanilla ice cream |
| --- | --- | --- | --- |
| c: spoiled cabbage | b: our son's birthday party | Px: x is a person | Lxy: x likes y |
| Cxy: x comes to y | Axy: x ate y | Syz: we should serve y at z | Txy: x throws away y |

## COMMUTATION (COM)

$$X \vee Y :: Y \vee X$$
$$X \& Y :: Y \& X$$

Someone likes vanilla ice cream, but nobody likes spoiled cabbage. If nobody likes spoiled cabbage, but at least one person likes ice-cream, then we shouldn't serve spoiled cabbage at our son's birthday party. So, we shouldn't serve spoiled cabbage.

1. (∃x)(Px & Lxi) & ~(∃x)(Px & Lxc)
2. [~(∃x)(Px & Lxc) & (∃x)(Px & Lxi)] → ~Scb         | ~Scb |
3. ~(∃x)(Px & Lxc) & (∃x)(Px & Lxi)         1, COM
4. ~Scb         2, 3 MP

Remember that *MP* states that if you have a conditional on one line and a statement that is *identical* to the antecedent of that conditional on another line, you can derive a statement which is identical to the consequent of the conditional. Whereas the statement on line 1 is *similar* to the antecedent of the conditional on line two, it's not *exactly* the same. The two conjuncts are "flipped" as it were. With the new rule, *Commutation*, we are able to use the statement on line 1 to derive a statement which *is* identical to the antecedent.

## ASSOCIATION (AS)

$$(X \vee Y) \vee Z :: X \vee (Y \vee Z)$$
**OR**
$$(X \& Y) \& Z :: X \& (Y \& Z)$$

Either Murray or Janice ate the vanilla ice cream or Lori threw it away. It's not the case that Janice ate it. Hence, either Murray ate it or Lori threw it away.

1. (Ami v Aji) v Tli
2. ~Aji                        ⬚ Ami v Tli ⬚
3. (Aji v Ami) v Tli     1, COM
4. Aji v (Ami v Tli)    3, AS
5. Ami v Tli                  2, 4 DS

Notice we used *Commutation* to derive a new statement on line #3 from the statement on line #1. We did this to allow use of *Association*. You can also see that we applied the rule to only part of the line, the part that was in the parentheses. Once we did that we were then able to apply *Association* which then let us use *Disjunctive Syllogism* to complete the proof. What we did on line #3 illustrates an important difference between implication and equivalence rules:

**Implication rules can be applied only to whole lines of a proof, not parts of a line. Equivalence rules can be applied to any part of a line.**

## DOUBLE NEGATION (DN)

$$X :: \sim\sim X$$

If Janice didn't throw away the spoiled cabbage, then we won't serve ice cream at the birthday party. Now, Janice threw away the spoiled cabbage only if she likes Lori. Fortunately, it turned out that we are going to be serving ice cream at the birthday party. So, it must be that Janice does like Lori.

1. ~Tjc → ~Sib
2. Tjc → Ljl
3. Sib                   ⬚ Ljl ⬚
4. ~~Sib            3, DN
5. ~~Tjc            1, 4 MT
6. Tjc                 5, DN
7. Ljl                  2, 6 MP

Double negation allows you to either add two tildes or remove two tildes to any statement. As you can see, we used DN both ways in the proof above. On line #3 we have a statement with no tildes, which allows us to derive the statement on line #4 that has two tildes. We needed it because the consequent of the conditional on line #1 is ~Sib. In order to use MT, we need to have the negation of that on a separate line, and the negation of ~Sib is *not* Sib, it is ~~Sib. We used DN again on line #5 where we have a statement that has two tildes and were then able to derive a statement on line #6 that has no tildes.

## CONTRAPOSITION (CONT)

$$X \to Y :: {\sim}Y \to {\sim}X$$

If Murray likes Janice, then Janice likes Lori. If Janice likes Lori, then she should come to the birthday party. The upshot is that if Janice doesn't come to the birthday party then Janice doesn't like Murray.

1. Lmj → Ljl
2. Ljl → Cjb          ~Cjb → ~Ljm
3. Lmj → Cjb     1, 2 HS
4. ~Cjb → ~Ljm   3, CONT

This rule can be a bit tricky as it applies to any conditional, including conditionals that have a tilde in one part and not in the other. However, this does allow you some flexibility. For example, if you had the conditional ~Kab → Rab and you applied CONT you would derive ~Rab → Kab OR ~Rab → ~~Kab.

## Exercise 12b

Construct proofs for the following arguments.

1. Ka v (Lt v St), ~(Ka v Lt)     St
2. (Ad v Bq) v Ct, ~Ad     Ct v Bq
3. ~(~Ba → An), Cab → (~An → Ba)     ~Cab
4. Bd → Ec, ~Fe v Gm, (Bd & Cm) & Da, (Da & Cm) → Fe     Ec & Gm
5. (Ga v Ha) → (Ja v Ka)     ~(Ja v Ka) → ~(Ha v Ga)
6. (Wp & ~Ca) v (Ka & Qd), (~Ca & Wp) → Ulm, (Ka & Qd) → Ta     Ulm v Ta
7. [~(Sd & Tb) v ~~Km] & (Tb & Sd)     Km
8. (Ab & Ca) → Ba, (~Ba v ~Ab) v Dab, Ca & Ab     ~~(Dab v Pab)
9. ~Ak & [(Ak v Bk) v Ct]     Bk v Ct
10. Fa → (Ga & Ha), (Ha & Ga) → Jm     Fa → Jm

## FOUR MORE EQUIVALENCE RULES

In this last section we will introduce the final four equivalence rules. This still won't give us a complete system, but we'll provide the last rules we need in the next chapter.

### TRANSLATION SCHEME
RD: Everything

| d: Dumbo | g: Goofy | f: Fido | Dx: x is a dog | Tx: x can talk |
|---|---|---|---|---|
| Ax: x is an animal | Px: x is a pig | Wx: x has wings | Ex: x is an elephant | Rx: x is real |

## BICONDITIONAL BREAKDOWN (BCB)

$$X \leftrightarrow Y :: (X \to Y) \& (Y \to X)$$

Goofy is a dog if and only if Fido is not a dog. If Goofy is a dog, then there are talking animals. So, if Fido is not a dog, then there are such things as talking animals.

1. Dg $\leftrightarrow$ ~Df
2. Dg $\to$ (∃x)(Ax & Tx)      ~Df $\to$ (∃x)(Ax & Tx)
3. (Dg $\to$ ~Df) & (~Df $\to$ Dg)    1, BCB
4. ~Df $\to$ Dg    3, SIMP
5. ~Df $\to$ (∃x)(Ax & Tx)    2, 4 HS

In the example above we only used the rule in one direction, going from a biconditional to a conjunction composed of two conditionals. Remember that if you want to go in the other direction, that is to derive a biconditional, you need to have *both* conditionals.

## MATERIAL IMPLICATION (MI)

$$\sim X \vee Y :: X \to Y \quad \text{(not ... unless rule)}$$

Either Dumbo's no elephant or pigs have wings. If pigs have wings, then Goofy can talk. If Goofy can talk, then talking animals don't exist. Clearly, either Dumbo's not an elephant or it's not the case that talking animals exist.

1. ~Ed v (x)(Px → Wx)
2. (x)(Px → Wx) → Tg
3. Tg → ~(∃x)(Ax & Tx)          | ~Ed v ~(∃x)(Ax & Tx) |
4. Ed → (x)(Px → Wx)            1, MI
5. Ed → Tg                       2, 4 HS
6. Ed → ~(∃x)(Ax & Tx)           3, 5 HS
7. ~Ed v ~(∃x)(Ax & Tx)          6, MI

Be careful when you use MI to pay close attention to the relationship between the conditional and the disjunction. Notice that it's the *antecedent* of the conditional that becomes the negated disjunct in your newly derived disjunction.

## DUPLICATION (DUP)

$$X :: X \vee X$$
**OR**
$$X :: X \& X$$

Either Dumbo is a real elephant or Goofy is real dog. If Goofy is a real dog, then there are talking animals. Of course, Dumbo is a real elephant only if there are talking animals. So, talking animals *do* exist.

1. (Rd & Ed) v (Rg & Dg)
2. (Rg & Dg) → (∃x)(Tx & Ax)
3. (Rd & Ed) → (∃x)(Tx & Ax)     | (∃x)(Tx & Ax) |
4. (∃x)(Tx & Ax) v (∃x)(Tx & Ax)  1, 2, 3 DIL
5. (∃x)(Tx & Ax)                  4, DUP

## DEMORGAN CRANK (DeM/CR)

~(X v Y) :: (~X & ~Y)   (neither ... nor rule)
~(X & Y) :: (~X v ~Y)   (not ... both rule)

Rather than immediately presenting an English language argument using this equivalency rule, we need to explain it a bit more fully. Whereas you can find DeMorgan presented in any number of logic and math books, the version of the rule we're using is slightly different. We'll start with the following statement: ~(X v ~Y).

Now think of inserting this statement into a machine with eight different slots and that each of those slots can be in one of two positions.

The first slot is either a tilde ~ or no tilde
The second slot is either a parenthesis ( or no parenthesis
The third slot is either a tilde ~ or no tilde
The fourth slot has only a statement X and no alternative
The fifth slot is either a wedge v or an ampersand &
The sixth slot is either a tilde ~ or no tilde
The seventh slot is a statement Y with no alternative
The eighth slot is either a parenthesis ) or no parenthesis

Now take the statement, ~(X v ~Y), put it in the machine and turn the crank.

And all the slots flip to:

This gives us ~X & Y.

Here's another example, ~X & ~Y. Picture it on the crank. Mentally turn the crank. What statement do you get?

You should have come up with: ~(X v Y).

As an equivalence rule, DeM/CR can be applied to part of a line, but this requires some careful attention. If you have a complex statement such as (~X v ~Y) & (~Y & X), consider all the various ways in which might apply DeM/CR. You could apply it to the left conjunct alone or to the right conjunct alone or to the conjunction as a whole. If we apply it to only part of a statement, we need to imagine that particular bit by itself, separated from the rest of the statement. For example, if we want to apply DeM/CR to only the left conjunct we then take that statement and look at it apart from the rest of the overall statement. If you took that left conjunct out it would look like this:

(~X v ~Y)

When we do this though we can see that the parentheses are superfluous, that is, if we took away the parentheses, we could still clearly identify the main operator. The parentheses are not essential to the nature of the statement. When you apply CRANK you need to imagine the statement you're applying it to *without* any superfluous parentheses. Take away the parentheses, then apply DeM/CR to *that* statement. So, in the above case, we'd actually be applying DeM/CR to ~X v ~Y, *not* to (~X v ~Y). When we do this we derive ~(X & Y). Having done this we can then take *that* statement and reinsert it into the original statement, thereby obtaining, ~(X & Y) & (~Y & X).

## Exercise 12c

Construct proofs for the following arguments.

1. ~(Ca & Da), (~Ca → Sb), ~Da → Td    [Sb v Td]
2. ~(~Ac v Bn)    [Ac]
3. Pa ↔ ~(x)Qx, ~(Pa v St)    [(x)Qx]
4. ~Pg, ~(Pg v Qa) → ~Rb, ~Qa    [~Rb]
5. ~Ea, Fs → (Da v Ea), ~Da    [~Fs]
6. ~Ab → ~~Rk, (z)Lz → ~Mn, ~Ab v (z)Lz    [~(~Rk & Mn)]
7. La → Ka, Ka → ~Cf, Cf v De, ~De    [~La]
8. ~Ia ↔ Ag, (Fn & Hj) v ~Ia, ~(Ag v Bn)    [Fn & ~Bn]
9. (Pd v Qb), Qb → St, Pd → (Ur v Ve), ~(St v Ve)    [Ur]
10. Rk → Bd, ~(Rk & ~Bd) → Jg    [Jg]
11. (Pa & Ga) → Re, (Re & Sd) → Ta, Pa & Sd, Ga v Re    [Re v Ta]
12. ~(Ca v Da), Da ↔ (Ep v Fm), ~Ak → (Ca v Fm)    [Ak]
13. ~(Se v Rb), Pn → Rb    [~Pn]
14. ~(Ja & Tf), ~(~Tf v Ea)    [~(Ja v Ea)]
15. Dar → ~Ba, ~Ca → Ba, ~Dar → ~Ca    [Dar ↔ Ca]

## CONDITIONAL PROOF AND REDUCTIO AD ABSURDUM

The last two rules are unique among the rules we use in that they require that we introduce a new kind of premise, one we'll call a *provisional assumption*. To see why we need this new rule consider the following argument:

If Bob wants to be a good person, then if he wants to be a lawyer he should be a public defender. So, if he wants to be a lawyer *and* he wants to be a good person, then he should be a public defender. (b: Bob; Lx: x wants to be a lawyer; Px: x wants to be a good person; Dx: x should be a public defender.)

Once symbolized the argument would look like this,

1. Pb → (Lb → Db)    [(Lb & Pb) → Db]

This is a valid argument, and the easiest way to demonstrate it is with the rule called *Conditional Proof* (CP). Conditional proof is aptly named since it is a rule that can be used whenever you want to derive a conditional statement. To understand the way *Conditional Proof* works, it helps to think about the nature of conditionals.

Remember that a conditional asserts a relationship *between* two statements. It says that **if** one statement (A) is true (the antecedent), then the other statement (C) is true (the consequent). So, if you can demonstrate that with one statement, *along with the other premises*, you can derive another statement, then you've shown

If you have A, then you can get C.

This is, in other words, the conditional: A → C. You, in effect, join the two statements into a new statement, a conditional.

Conditional proofs require us to introduce a new concept, a *provisional assumption*. A provisional assumption is a statement that you introduce into the argument that will function as a premise in a mini argument that is designed to derive another statement. When you introduce the provisional assumption you write "PA" off to the right, where you normally cite the appropriate rule for that particular line. You then use the rules covered in the previous sections of the chapter to derive the desired statement. For CP, the desired statement is the consequent of the conditional. When you have derived the statement you'll have completed the "mini argument." Having completed the mini argument, you will have derived a new statement, a conditional that has the provisional assumption as its antecedent and the conclusion of the mini argument as its consequent.

In order to clearly identify what part of the argument constitutes the mini argument, you *create a box around the section of the proof that is the mini proof*. You begin the box when you introduce the provisional assumption, which in this case is, "Lb & Pb." Having introduced the provisional assumption, we will then use it plus the premises we started with and the rules to derive the consequent of the statement we're after, which in this case is "Db"

1. Pb → (Lb → Db)            | (Lb & Pb) → Db |
2. Lb & Pb            PA
3. Pb                 2, SIMP
4. Lb → Db            1, 3 MP
5. Lb                 2, SIMP
6. Db                 4, 5 MP

Once the consequent is derived, we close the box around the mini-argument.

1. Pb → (Lb → Db)            | (Lb & Pb) → Db |
2. Lb & Pb            PA
3. Pb                 2, SIMP
4. Lb → Db            1, 3 MP
5. Lb                 2, SIMP
6. Db                 4, 5 MP

The box serves as a clear indicator of the *scope* of the mini-argument, and hence, what exactly will be the antecedent and what will be the consequent of the new conditional. The antecedent is the *first* line of the box, and the consequent is the *last* line of the box.

On the line immediately following the box, we write the new statement and cite as its justification the line that starts the box (in this case it's line #2) *through* the last line of the box (in this case, line #6). After the mini proof is completed, the overall proof would look like this,

1. Pb → (Lb → Db)
2. Lb & Pb          PA
3. Pb               2, SIMP
4. Lb → Db          1, 3 MP
5. Lb               2, SIMP
6. Db               4, 5 MP
7. (Lb & Pb) → Db   2 – 6, CP

For this particular argument the proof is finished.

At this point we need to spell out some additional aspects of this rule. First, it is important to recognize that the provisional assumption functions only *somewhat* like the premise of an argument; there are limitations on the way it can be used. For example, when you introduce a statement as a provisional assumption, that statement *must* eventually become the antecedent of a conditional. When the conditional is finally derived, we say the provisional assumption has been *discharged*, and a proof is not completed until this has taken place. *When a PA is discharged that means nothing depends on that premise being true!* Think about it. In the CP, above, the conclusion **only** says that **IF** (Lb & Pb) **were** true, then Db is also true. It **doesn't** say that (Lb & Pb) **is in fact** true.

Another point to note is that *once the box is closed, nothing within the box can be used in any other part of the proof.* In this way, the box around the mini-argument is like the box that surrounds the conclusion: **what's in the box stays in the box**. It can't be repeated, or used in any rule to get a statement, outside the box.

Finally, just because this particular proof is done after the completion of the mini-argument, doesn't mean that proofs always have to finish at that point. Having completed the mini-proof, the proof can continue on for several lines if need be. What is more, it is possible to construct mini-proofs *within* other mini-proofs. We will return to this last point after we have introduced our next rule.

> Once mini-argument is complete and box is closed, we can't use any line *within* the box farther down in the proof.

Conditional proofs can also be used when what you're aiming at is not a conditional. CP is a useful technique for conclusions that are disjunctions or negated conjunctions. We'll briefly show examples of each of these proofs.

For the argument

$$(E \vee F) \rightarrow {\sim}A, \quad {\sim}F \rightarrow B \quad \boxed{{\sim}A \vee B}$$

You now know by the rule Material Implication (MI) that our conclusion is equivalent to the conditional A → B. So for this proof:

Assume the antecedent, **A**
Work for the consequent, **B**
Derive the conditional, **A → B**
Use MI to get the conclusion, **~A v B**

Here's the complete proof:

1. (E v F) → ~A
2. ~F → B                $\boxed{{\sim}A \vee B}$

| | |
|---|---|
| 3. A | PA |
| 4. ~~A | 3, DN |
| 5. ~(E v F) | 1, 4 MT |
| 6. ~E & ~F | 5, DeM/CR |
| 7. ~F | 6, SIMP |
| 8. B | 2, 7 MP |

9. A → B         3 – 8, CP
10. ~A v B        9, MI

Now for a negated conjunction. Look at the following argument, think about the equivalence rules that would let you express the conclusion as a conditional, and how you would construct a CP proof for that conditional. After you've done this, look carefully at the completed proof.

The argument: **~Prt v (Cb → Dtr), (~Cb v Dtr) → ~Qb**     $\boxed{{\sim}(Prt \,\&\, Qb)}$

1. ~Prt v (Cb → Dtr)
2. (~Cb v Dtr) → ~Qb      $\boxed{{\sim}(Prt \,\&\, Qb)}$

| | |
|---|---|
| 3. Prt | PA |
| 4. ~~Prt | 3, DN |
| 5. Cb → Dtr | 1, 4 DS |
| 6. ~Cb v Dtr | 5, MI |
| 7. ~Qb | 2, 6 MP |

8. Prt → ~Qb       3 – 7, CP
9. ~Prt v ~Qb      8, MI
10. ~(Prt & Qb)     9, DeM/CR

The last rule we will introduce in this chapter is known as *Reductio Ad Absurdum (RAA)*. It is similar to conditional proof in that it too employs a provisional assumption, a mini-argument, and a box. This rule depends upon a concept we have used in previous chapters, the contradiction. Recall that a contradiction is when you have two statements, one of which is the denial of the other. On a truth tree, a contradiction closes the branch. Something similar happens in the RAA, the "argument to absurdity" proof. The idea behind RAA is that if you have a statement that can be shown to lead to a contradiction, an absurd state of affairs, then you have demonstrated that *that statement must be false*. We'll demonstrate this new rule with the following argument:

If Homer is arrested, then he'll need to post bail. Therefore, it's not the case that Homer can be arrested and not have to post bail. (h: Homer; Ax: x is arrested; Bx: x needs to post bail.)

Symbolized, it looks like:

$$1.\ Ah \rightarrow Bh \quad \boxed{\sim(Ah\ \&\ \sim Bh)}$$

RAA begins in a similar fashion to CP; you introduce a provisional assumption and start a mini-argument. In the case of RAA, though, you're not creating a conditional, and so you need to make a different choice as to what you introduce as the provisional assumption. *For RAA you introduce a statement that is the negation of the statement you want to derive.* The goal of the mini argument is to create a contradiction, *any* contradiction. We'll introduce the negation of the conclusion as the provisional assumption, and then attempt to derive a contradiction within the mini-argument.

| | | |
|---|---|---|
| 1. Ah → Bh | | ~(Ah & ~Bh) |
| 2. Ah & ~Bh | PA | |
| 3. Ah | 2, SIMP | |
| 4. ~Bh | 2, SIMP | |
| 5. ~Ah | 1, 4 MT | |
| 6. Ah & ~Ah | 3, 5 CONJ | |

We have derived a contradiction at line #6, which would not be possible if all our premises are true. Since what we want to know is whether our original argument is valid, that is, if the first premise were true, it would imply the conclusion, we "hold" that the premise on line #1 is true. That can only mean our provisional assumption must be false. This allows us to derive a new statement on line #7, which is the negation of PA. Just as was the case with CP, we enclose the mini-proof in a box, justify the new statement by referencing the first through last lines of the box and cite the rule. The final proof looks like this,

| 1. Ah → Bh | | ~(Ah & ~Bh) |
|---|---|---|
| 2. Ah & ~Bh | PA | |
| 3. Ah | 2, SIMP | |
| 4. ~Bh | 2, SIMP | |
| 5. ~Ah | 1, 4 MT | |
| 6. Ah & ~Ah | 3, 5 CONJ | |
| 7. ~(Ah & ~Bh) | 2 – 6, RAA | |

As was the case with CP, what's in the box stays in the box. The statement on line #7 could be used if you did need to continue the proof, but lines 2 through 6 can't be used in any other part of the proof.

Before we conclude this chapter we need to return to a point raised earlier. As we pointed out in our discussion of CP, it is possible to have mini-arguments within mini-arguments. We'll finish the chapter by demonstrating two such arguments.

Note below that the mini-arguments of the outer boxes do not reference any lines in the inner boxes. They reference the *conclusions* of the mini-arguments, but not the lines that lead up to those conclusions. For greater clarity we indent each time we introduce a new PA, but you don't need to do this when constructing your proofs. Just clearly boxing each mini-argument will show the extent of each mini-argument.

*Argument 1*: If it's false that Sam is a college graduate, then Frank will fire him only if he lied about it. Of course, if Frank fires him then he will be prosecuted if he lied about it. So, if it's false that Sam is a college graduate, then if Frank fires him, he's going to be prosecuted. (s: Sam; f: Frank; Lx: x lies about graduating college; Cx: x is a college graduate; Px: x is prosecuted; Fxy: x fires y.)

| 1. ~Cs → (Ffs → Ls) | | ~Cs → (Ffs → Ps) |
|---|---|---|
| 2. Ffs → (Ls → Ps) | | |
| 3. ~Cs | PA | |
| 4. Ffs | PA | |
| 5. Ls → Ps | 2, 4 MP | |
| 6. Ffs → Ls | 1, 3 MP | |
| 7. Ls | 4, 6 MP | |
| 8. Ps | 5, 7 MP | |
| 9. Ffs → Ps | 4 – 8, CP | |
| 10. ~Cs → (Ffs → Ps) | 3 – 9, CP | |

*Argument 2*: If John wants to go to New York then George will give him a lift, only if George's car is working. We can conclude that if John wants to go to New York then if George gives him a lift, then his car is working. (j: John; g: George; Nx: x wants to go to New York; Lxy: x gives y a lift; Cx: x's car is working.)

1. (Nj → Lgj) → Cg         Nj → (Lgj → Cg)
2. Lgj                      PA
3. ~Cg                      PA
4. ~(Nj → Lgj)              1, 3 MT
5. ~(~Nj v Lgj)             4, MI
6. Nj & ~Lgj                5, DeM/CR
7. ~Lgj                     6, SIMP
8. Lgj & ~Lgj               2, 7 CONJ
9. Cg                       3 – 8 RAA
10. Lgj → Cg                2 – 9 CP
11. ~Ng v (Lgj → Cg)        10, ADD
12. Nj → (Lgj → Cg)         11, MI

## STRATEGY TIPS

While you now have most of the rules, it can sometimes be a bit daunting in trying to figure out how to proceed. It's one thing to know that *Modus Ponens* is an acceptable move in a proof in front of you and it's another to know whether it's a good idea to use it at that moment or to use a different, equally acceptable rule. With this in mind we've gathered together a few strategy tips that we hope will help you when you're stuck.

**Do Something!**: When you're stuck, just do something. Apply any rule that you can legitimately use, even if you don't know where you're going with it.

**Conclusion in the Premises**: Look at the conclusion and see if there is anything like it in the premises.

**Mini-Goals**: Set yourself a series of mini-goals, designed to lead from one to the other until it leads to the conclusion.

**Build a Conjunction**: If your conclusion is a conjunction, work first for one of the conjuncts and then the other, then use CONJ to put the two together.

**Disjunction/Addition**: If your conclusion is a disjunction, try obtaining one of the disjuncts and then use ADD to bring in the other disjunct.

**Backward from the Conclusion**: Consider what the conclusion would look like if you applied an equivalence rule to it.

**CP for Conditional Conclusions**: If your conclusion is a conditional, immediately consider CP as a strategy. Assume the antecedent and try working for the consequent.

**Negated Conclusions**: If you have a negated statement as a conclusion, try RAA. Make a provisional assumption of the statement without its tilde and start trying to generate a contradiction.

## Exercise 12d

Construct either a CP or an RAA proof for each of the following.

1. ~(Rm & ~Ta)   $\boxed{Rm \to Ta}$
2. Rab → Tab   $\boxed{\sim(Rab\ \&\ \sim Tab)}$
3. (Mz → Sd) & (~Mz → Sd)   $\boxed{Sd}$
4. Am & ~Am   $\boxed{De}$
5. ~Kb   $\boxed{Kb \to Za}$
6. Pa → Qm, ~Pa → Ja, ~Qm → ~Ja   $\boxed{Qm}$
7. (x)Zxa → Gl, (x)Zxa v Gl   $\boxed{Gl}$
8. [(∃x)Mx → (∃y)Ky] → (∃x)Mx   $\boxed{(\exists x)Mx}$
9. ~Ac & ~Ba   $\boxed{Ac \leftrightarrow Ba}$
10. ~Pr → (Ra & Sa), ~Qr → (Ra & Tr), ~(Sa v Tr)   $\boxed{Pr\ \&\ Qr}$
11. (Am → Ba) → Cd, Am → ~(Eq v Fq), Eq v Ba   $\boxed{Am \to Cd}$
12. ~Jk v Lf, Jk v ~Lf   $\boxed{(Jk \leftrightarrow Lf)}$
13. (~Mo v ~Bg) → ~Cg   $\boxed{Cg \to Mo}$
14. (Zd & Hm) → ~Fp, Fp v (Gp v Wr), Zd → Hm   $\boxed{Zd \to Gp}$
15. (Kc v Na) → ~Ce, Dh → (~Fab & ~Gj)   $\boxed{(Kc\ v\ Dh) \to \sim(Ce\ \&\ Fab)}$
16. (Mb & Rt) v (~Mb & ~Rt)   $\boxed{Mb \leftrightarrow Rt}$
17. (Ak v Ql) → ~(Gd & Dj), (Ak & Dj), ~Gd → ~(Ca & Dj)   $\boxed{\sim(Ak\ \&\ Ca)}$
18. Cnf → (Dt → ~Cnf), Cnf ↔ Dt   $\boxed{\sim Cnf\ \&\ \sim Dt}$
19. Abg ↔ ~(Bh v Cs), Bh ↔ (Ds & ~Epo), ~(Epo & Abg)   $\boxed{Abg \to \sim Ds}$
20. Wt → Xp, (Wt → Yk) → (Hn v Xp), (Wt & Xp) → Yk, ~Hn   $\boxed{Xp}$

# Chapter Thirteen: Proof Rules for Quantifiers

*What's Up?*
UE, EC, EE, UC
QN
II, Sym, IR

In this chapter we conclude the method of proof by introducing rules that allow you to construct proofs for arguments that involve statements with quantifiers or identity statements. There will be eight rules in all, five for quantified statements and three for identity statements.

## UNIVERSAL ELIMINATION (UE)

Our first rule is called *Universal Elimination* and it allows us to derive a non-universal statement from a universal. The basic idea behind it is quite simple. Consider the following universal statements,

**TRANSLATION SCHEME**
RD: Everything

| o: the oak in our front yard | a: my aunt's fern | s: Susan | c: the cement truck |
|---|---|---|---|
| Tx: x is a tree | Px: x is a plant | Lx: x is living | Fxy: x feeds off y |

1. Everything is alive.
2. All trees are alive.
3. All trees are plants.
4. Everything feeds off something.
5. Everything is not alive.
6. Everything feeds off of Susan.

Translated, the statements read as follows:

1. (x)Lx
2. (x)(Tx → Lx)
3. (x)(Tx → Px)
4. (x)(∃y)Fxy
5. (x)~Lx
6. (x)Fxs

Because these are universal statements, they're asserting that some property obtains for everything in our range of discourse (as in statements 1 and 5) or all the members of some *class* of things in our range of discourse (like the class of "trees" in 2 and 3). They may be simply asserting that some property applies (as in 1 or 5) or they may be claiming a relationship between things (as in 2, 3, 4, and 6). Whatever kind it is, a universal claim has implications beyond itself.

For example, consider statement #1, "Everything is alive" and its relationship to the following two statements:

a. My aunt's fern is alive.
b. The cement truck is alive.

Both "a" and "b" are implied by statement #1. After all, if some claim is true of everything (in the range of discourse), then no matter what we pick (from within that range) it will be true of it as well.

Or consider statement #3 "All trees are plants" and its relationship to the following two statements:

a. If the oak in our front yard is a tree, then it's a plant.
b. If Susan is a tree, then it's a plant.

Statement #3 is asserting that there is a universal relationship between something being a tree and it being a plant. You may have to establish that it's a tree first, but once you've shown that, statement #3 allows you to automatically conclude that it's also a plant.

What all this shows is that we can have a rule that says when we have a universal statement we can derive a new statement on another line in which we *eliminate the universal quantifier and replace all the variables that are bound to it with <u>any</u> individual constant*. We can represent the rule as follows:

**Universal Elimination (UE):**
(x)Px   | Pa |

We'll show this in the following argument.

ARGUMENT #1: Everything's alive. So, my aunt's fern is alive and the oak tree in my front yard is alive.

Once symbolized the argument would be

1. (x)Lx        |La & Lo|

We implement it exactly the same way we do all our earlier rules in that you write out the new statement, give the line number that you're using as the basis for your claim, and the rule that's letting you derive the new statement. The proof for it would be as follows:

1. (x)Lx       |La & Lo|
2. La        1, UE
3. Lo        1, UE
4. La & Lo    2, 3 CONJ

Notice that it's possible to derive more than one statement from the same universal statement. You can go back to the well, so to speak, as many times as you like.

Notice too that UE is *just like our tree rule* for instantiating universals. Just as our tree rule allows us to choose any constant we need in our attempt to finish the tree, UE allows us to choose any constant we need to finish the proof. And, again, just as our tree rule allows us to instantiate a universal as often as we need, so too does our UE proof rule.

Now consider this argument,

ARGUMENT #2: All trees are alive. The oak in my front yard is a tree, and of course, the cement truck is not alive. So, the oak is alive and the cement truck isn't a tree.

Symbolized, the argument looks like this,

1. (x)(Tx → Lx)
2. To & ~Lc    |Lo & ~Tc|

Since the conclusion is a conjunction, our usual strategy is to work for one of the conjuncts, and having secured one, work for the other. Once we have both conjuncts on different lines, we use *CONJ* to bring them together into a single conjunction. It would seem as though we have everything we need to get started on line #2. However, while it's tempting to use *Simplification* on line #2 to separate the two statements and then begin using *Modus Ponens* and *Modus Tollens* to generate our conclusion, there are two reasons we can't.

First, both *Modus Ponens* and *Modus Tollens* require a conditional to work and the statement on line #1 is not a conditional, it's a universal statement. This reminds us again, why it's so important to be able to identify the main operator, which in this case is a universal quantifier. The second reason we can't just start using our *MP* or *MT* is that neither of the two statements on line #2 is what we would need to use those rules. *MP* requires a statement that is identical to the antecedent of a conditional, and we don't have that on line #2. *MT* requires a statement which is identical to the negation of the consequent of a conditional, and that's not on line #2 either.

However, we can use UE to generate what we need. First, we will use UE to derive a conditional that lets us utilize MP.

1. (x)(Tx → Lx)
2. To & ~Lc          | Lo & ~Tc |
3. To → Lo           1, UE
4. To                2, SIMP
5. Lo                3, 4 MP

Having secured one of the conjuncts of the conclusion, "Lo" we now can turn to the other "~Tc." To derive this we need a conditional again, although we can't use the conditional on line #3 since it has "o" as its individual constant, and we need "c." Fortunately, as we demonstrated earlier, you can go back to the universal statement as often as you like. So, we'll use UE again, only this time we will derive a conditional that has "c" as its individual constant. The final proof would thus look like,

1. (x)(Tx → Lx)
2. To & ~Lc          | Lo & ~Tc |
3. To → Lo           1, UE
4. To                2, SIMP
5. Lo                3, 4 MP
6. Tc → Lc           1, UE
7. ~Ac               2, SIMP
8. ~Tc               6, 7 MT
9. Lo & ~Tc          7, 8 CONJ

This proof illustrates another important point about UE. First, when you use UE *all* the variables that are bound to the quantifier have to be replaced with the *same* letter. We couldn't have used UE to generate "To → Lc" on line #3 or #6.

## EXISTENTIAL CREATION (EC)

As we saw with trees, our quantifier rules allowed us to "get rid" of quantifiers—to create instances of universal or existential quantifications. Our next proof rule, however, allows us to *introduce* an existential quantifier. To see how this rule works consider the following statements.

**TRANSLATION SCHEME**
RD: Everything

| d: Detroit | m: Miami | n: Newark | s: Sam |
| --- | --- | --- | --- |
| a: Allen | b: Bob | Px: x is a person | Gx: x is great |
| Lxy: x is larger than y | Ixy: x is in y | Exy: x is further east than y | Cxy: x is colder than y |

1. Detroit is great.
2. Bob is great.
3. Detroit is further east than Newark.
4. Sam is a person and so is Bob.
5. Sam is larger than someone.
6. Something is colder than Detroit.
7. Everything is colder than Miami.

Symbolized the statements would read

1. Gd
2. Gb
3. Edn
4. Ps & Pb
5. (∃x)(Px & Lsx)
6. (∃x)Cxd
7. (x)Cxm

The new rule, Existential Creation (EC) allows us to move from a statement with an individual constant to an existential statement; in other words, it allows us to introduce an existential quantifier into a proof.

This rule operates on the following simple idea that we can best illustrate by considering sentences #1 and #2. If it's true that Detroit is great, then we can certainly conclude that *something* is great. If it's true that Bob is great, then we can certainly conclude that *something* is great. The rule therefore allows us to replace an individual constant with a variable and bind that variable to an existential quantifier.

The rule can be represented as follows

**Existential Creation (EC):**
Pa      (∃x)Px

This new rule allows us to construct a proof for the following argument.

ARGUMENT #3: Bob's in Detroit and Detroit's further east than Miami. If something is in Detroit, then if Detroit is further east than Miami, then Bob is further east than Miami. So, something is further east than Miami.

1. Ibd & Edm
2. (∃x)Ixd → (Edm → Ebm)   (∃y)Eym
3. Ibd         1, SIMP
4. (∃x)Ixd     3, EC
5. Edm → Ebm   2, 3 MP
6. Edm         1, SIMP
7. Ebm         5, 6 MP
8. (∃y)Eym     7, EC

Be careful in how you use this rule. The following for example would *not* be allowed,

1. Ps & Bs
2. (∃x)(Px & B**y**)   1, EC      MISTAKE!

This is wrong because if you're going to change the constants to variables, they have to be the *same* all the way across. It's permissible to change only one of them if you want and to leave the remaining one as an individual constant. However, it's not permissible to alter them to different variables.

1. Ps & Pb
2. (∃x)Px & Pb   1, EC      MISTAKE!

This is also an incorrect application of the rule EC. As we said in applying the rule, it must generate an existential quantifier, and in this case the statement on line #2 is a conjunction. Another way of saying this is that the rule cannot be applied to *part* of a line. Of course, you could use SIMP to derive Ps and Pb on two separate lines, and *then* apply EC to just Ps and then use CONJ to put the statements together.

Another potential mistake is the following

1. Ps & Mb
2. (∃x)(Px & Mx)   1, EC      MISTAKE!

This too is an incorrect application of the rule. In this case, the problem is that we've introduced an existential quantification that essentially claims that there exists at least one entity with both these predicates. But premise 1 may be about *two different* individuals—unless b = s, but we're not told that's true. So it's not legitimate to make the much stronger claim that *one* person has both these qualities. After all, 1 might be true because s is Sally, and Sally is P (visiting Paris), but b is Barry, and Barry is M (stuck in Montana).

Another potential mistake would be,

1. (x)Pxa
2. (∃x)(x)Pxx    1, EC    MISTAKE!

Here, in selecting x to replace a, we chose a variable that was *already* bound, which is not acceptable under EC. It would have been okay to pick u, w, v, y, or z, just not x.

To better see the various ways this rule *can* be used, we'll give you three examples of a proper use of EC.

1. Gd & Md
2. (∃x)(Gx & Mx)    1, EC

1. (x)Pxa
2. (∃y)(x)Pxy    1, EC

1. Gd & Gd
2. (∃x)(Gx & Gd)    1, EC

## Exercise 13a

Construct proofs for the following arguments. As we did before, we'll present the arguments in the Horizontal Version, and that means that you'll need to first put the arguments into Vertical Version to start the proof.

1. (x)(y)(z)(Mxy → Myz), Mlj    Mjl
2. ~Kaa    (∃y)~Kay
3. (∃x)~Rxx, (∃x)~Rxx → (x)(y)~Txy    ~Tba
4. (y)(Bay v Bya)    Baa
5. Meb → Nbb, (z)~Nzz    ~Meb
6. (x)(y)(z)[(Wxy & Wyz) → Wxz], Wab, Wbc    Wac
7. (x)(Tx → (y)[Vy → (∃z)(Pz & Syz)]), Tb & Ve    (∃z)(Pz & Sez)
8. ~(x)(y)Pxy, ~Dab → (x)(y)Pxy    (∃x)(∃y)Dxy
9. (y)(Oyd → Cy), Ocd & Pcd, (z)(~Pzd v Lz)    (∃x)(Lx & Cx)
10. ~[~(x)Fx & Gbf], ~Gbf → (∃x)Hx, (x)Fx → Pf, ~(∃x)Hx    Mg → (∃y)Py

## EXISTENTIAL ELIMINATION (EE)

As the name suggests, Existential Elimination (EE) allows us to remove an existential quantifier and replace all the variables bound to it with some individual constant.

As we saw in our chapter on truth trees and quantifiers, we can replace an existential quantification by creating an instance of the quantification, so long as certain conditions obtain. Consider the following statement,

$$(\exists x)Px \quad (Px: x \text{ is a person})$$

This says that there exists something that is a person. Of course this is much different than what a universal quantifier would assert, namely that *everything* is a person. With a universal statement we'd have no problem removing the quantifier and replacing it with any name we wanted, since if everything is a person, then any individual we select would be a person. However, if we are to do the same thing with this existentially quantified statement, we would need to select an individual constant quite carefully.

For example, if something is a person, then we could put a name to this thing just so long as we make sure that the name we select is random, that it's arbitrarily chosen. As we said in the chapter on truth trees, it's like saying, "Something is a person, let's just call it Bob." We can be certain that we have arbitrarily selected the name if we obey the following two rules.

1. We cannot select any individual constant that has appeared previously in the proof
2. We cannot select any individual constant that appears in the conclusion of the proof.

These rules are in addition to the ones we introduced with regard to UE and EC: whatever individual constants you select must be replaced consistently and you can't quantify different names with the same quantifier.

To better understand the rule consider how it would be used in the following argument,

ARGUMENT #4: All gophers are mammals. Furthermore, some gophers aren't friendly. So, we can conclude that some mammals aren't friendly.

1. (x)(Gx → Mx)
2. (∃x)(~Fx & Gx)         |(∃x)(~Fx & Mx)|
3. ~Fa & Ga         2, EE ←        Tip: EE before UE (just like trees!)
4. Ga → Ma         1, UE ←

5. Ga          3, SIMP
6. Ma          4, 5 MP
7. ~Fa         3, SIMP
8. ~Fa & Ma    6, 7 CONJ
9. (∃x)(~Fx & Mx)    8, EC

On line #3, we selected the letter "a," since it doesn't show up in any previous line of the proof, thereby meeting our first condition. Then we check the conclusion of the proof to make sure that the individual constant we selected for EE did not occur there, thereby meeting the second requirement.

We can now represent the rule *Existential Elimination* as follows,

**Existential Elimination (EE):**
(∃x)Px    Pa

Provided that the individual constant selected has not been mentioned previously in the proof and does not appear in the conclusion of the proof.

## UNIVERSAL CREATION (UC) AND QUANTIFIER NEGATION (QN)

Our next rule allows us to create a universal statement from a statement about an individual. If we have a statement of the following sort, "Pa v Ga" we can then use the new rule, UC to derive a statement such as,

(x)(Px v Gx)
(y)(Py v Gy)
(z)(Pz v Gz)

In general:

**Universal Creation (UC):**
Pa    (x)Px

Now this rule may seem problematic. If you have the statement, "My cat has fleas," or Fc, it is *illogical*, from that *one* piece of information, to infer that "Everything has fleas," (x)Fx. Or if I knew that *something* has fleas [(∃x)Fx], it is illogical to conclude that *everything* has fleas, (x)Fx.

So, our UC rule applies in *certain restricted circumstances only*. Consider the following two arguments.

> UC Upshot:
> Universals are only created from other universals!

ARGUMENT #5: Everything is material. It also happens to be the case that everything expends energy. This means that everything is both material and an energy expender.

ARGUMENT #6: All material things expend energy. Of course, everything in existence is material. Given that we can conclude that everything expends energy.

These are both valid arguments, but with our current set of rules it's hard to show this. See how we use UE to prove Arguments 5 and 6 below.

| | | | | | |
|---|---|---|---|---|---|
| 1. (x)Mx | | | 1. (x)(Mx → Ex) | | |
| 2. (z)Ez | (y)(My & Ey) | | 2. (w)Mw | (y)Ey | |
| 3. Ma | 1, UE | | 3. Ma → Ea | 1, UE | |
| 4. Ea | 2, UE | | 4. Ma | 2, UE | |
| 5. Ma & Ea | 3, 4 CONJ | | 5. Ea | 3, 4 MP | |
| 6. (y)(My & Ey) | 5, UC | | 6. (y)Ey | 5, UC | |

The restriction on UC is this: *UC is used on a statement about individuals, ONLY IF that statement was itself derived from universal statements earlier in the proof.*

This restriction on UC rules out those cases of generalizing from my cat having fleas to everything having fleas. The idea behind this restriction is that you cannot apply UC to a statement if it is derived from a single instance. Here are some examples of "UC gone wrong."

1. Fc            (x)Fx
2. (x)Fx        1, UC        MISTAKE!

1. (x)(Px → Ixb)
2. Pa            (x)Ixb
3. Pa → Iab    1, UE
4. Iab           2, 3 MP
5. (x)Ixb       4, UC        MISTAKE!

1. Sd & (x)Sx    (x)Sx
2. Sd             1, SIMP
3. (x)Sx         2, UC        MISTAKE! ←

Note: (x)Sx is a valid conclusion from this premise, but not the way we did it. The correct way to derive this conclusion, of course, is to use SIMP to get (x)Sx directly from line 1.

In all these examples, what is happening is that we're trying to generalize to everything from what we know about one particular thing.

A particular sort of move that's ruled out by the restriction is this: *UC cannot be applied to a name that was derived from the implementation of Existential Elimination.* So, the following would not be an acceptable application of the rule.

1. (∃x)Px
2. Pa           1, EE
3. (z)Pz        2, UC        MISTAKE!

Remember that EE just introduces a name at random: "We'll call him Bob." The only thing we know about "Bob" is that he has this one property. That being the case, we can't validly infer that everything shares this property with "Bob."

Finally, just as in the previous cases, be sure you're using UC correctly in terms of the scope of the operator. The resulting statement from a use of UC has to be a universal quantification. This would not be valid,

1. Pa & Ma
2. (y)Py & Ma   1, UC        MISTAKE!

The statement derived here is a *conjunction*, not a universal quantification, and UC can only create universals.

Here is an argument that uses this new rule correctly:

ARGUMENT #6: No dogs are reptiles. All Scotties are dogs. So, no Scotties are reptiles. (Dx: x is a dog; Sx: x is a Scotty; Rx: x is a reptile.)

1. (x)(Dx → ~Rx)
2. (x)(Sx → Dx)    (x)(Sx → ~Rx)
3. Da → ~Ra        1, UE
4. Sa → Da         2, UE
5. Sa → ~Ra        3, 4 HS
6. (x)(Sx → ~Rx)   5, UC

Again this example illustrates our restriction: *UC is used on a statement about individuals ONLY IF that statement was itself derived from universal statements earlier in the proof.*

The last rule we'll introduce in this section is an equivalency rule called **Quantifier Negation (QN)**. This rule is similar to the tree rule that we used for negated quantifiers that occur on a tree. As we saw earlier, a statement like the following, "Nothing is free" (Fx: x is free) can be symbolized in one of two different ways. One way would be,

~(∃x)Fx   (There doesn't exist one thing that's free.)

and the other one is,

(x)~Fx   (Everything is non-free.)

Since these are both equally acceptable symbolizations of the statement we are able to have a rule that allows us to move between the two different symbolizations. Remember with the tree rule we only went "one way," from negated quantifier to quantification. This rule allows us to go "both ways." Note that we turn the first version into the second by moving the tilde in and changing from an existential quantifier to a universal quantifier. We could have just as easily gone in the opposite direction. Anytime we have a negation coupled with a quantifier we can move the negation in, or out, and change the quantifier. We can represent the rule as follows:

**Quantifier Negation (QN)**
~(x)Px :: (∃x)~Px
**OR**
(x)~Px :: ~(∃x)Px

And again, since this is an equivalence rule, it can be applied to any part of a line, as well as to whole lines, as illustrated below.

1. (x)~(Px → Lx)              1. (∃x)~(Pa & La) v (y)Ky
2. ~(∃x)(Px → Lx)   1, QN    2. ~(x)(Pa & La) v (y)Ky   1, QN

## Exercise 13b

Construct proofs for the arguments below.

1. (x)(Ax & Bx)                                              | (x)Ax & (x)Bx
2. ~(x)~Ax, (∃x)Ax → (x)Bx, (x)[(Bx v Cx) → Dx]              | (x)Dx
3. ~(∃x)(Rx v Sx) v (x)Tx, (∃x)~Tx                           | ~(∃x)Sx
4. (x)Ax → (x)Bx, ~(x)Bx                                     | (∃x)~Ax
5. ~(∃y)Cy, (y)~Cy → (z)Dz                                   | Db
6. ~(x)~Fx                                                   | (∃x)Fx
7. (∃y)Hy → (∃y)Jy, (y)~Jy                                   | ~Ha
8. ~(∃x)~Px, ~(∃y)Sy v ~(x)Px                                | ~Sd
9. (x)~Kx v (∃x)Lx, (x)~Lx                                   | ~Kb
10. (∃x)[Kx & (Ax & Sx)],(x)[(Ax v Sx) → ~Wx]                | (∃x)(Kx & ~Wx)

## IDENTITY RULES

The method of proof can extend to arguments containing identity statements, but it requires three new rules.

### TRANSLATION SCHEME
RD: People

| a: Anakin Skywalker | d: Darth Vader | l: Luke Skywalker | Bx: x is a brilliant fighter |
| Wx: x is a warrior | Px: x is peaceful | Gx: x is great | |

We will begin with a somewhat unusual rule, but one that's clearly valid. A truth of logic is that anything is identical to itself. The rules of derivation are designed to guarantee that each step is a true statement given the truth of the premises, and truths of logic are guaranteed true, period. Hence, we can introduce as a line on a proof a statement of the form, a=a, *any time* in a proof.

This relies not on any previous line of the proof, but just on the truth of logic that's everything is identical with itself. In our justification column for this line we'd just write II, for Identify Introduction, without any line number accompanying it. This is the *only* rule in our system that doesn't require a line number reference.

### Identity Introduction (II):
**a=a**
(where 'a' can be any constant)

All this says is that "Anakin is identical to Anakin." Notice that this doesn't require us to accept that Anakin exists or that he doesn't exist. Whether Anakin existed or he didn't, it would still be the case that Anakin is identical to Anakin.

The second rule can be illustrated with the following statement,

Darth Vader is actually Anakin Skywalker.

that could be symbolized as,

d=a

It would of course have been just as acceptable to symbolize the statement as

a=d

The same would be true if we said that Darth Vader is not Anakin Skywalker. This could be symbolized as either,

$$\sim a=d \quad \text{or} \quad \sim d=a$$

This simply means that any time in a proof that you have an identity statement, you can switch the individual constants around. This gives us the equivalency rule,

**Symmetry (SYM):**
a=b :: b=a
OR
~a=b :: ~b=a

Finally, the last identity rule is an implication rule called **Identity Replacement (IR)**. This is like the tree identity rule we've already seen. Consider the following argument.

ARGUMENT #7: Anakin Skywalker is a brilliant fighter. Anakin and Darth Vader are the same person. Hence, we can conclude that Darth Vader is a brilliant fighter.

1. Ba
2. a=d      | Bd |

**Identity Replacement** allows us to exchange individual constants that are part of a predicated statement if it's been established that two individuals are identical to each other. Notice that this is the only rule introduced in this chapter that requires *two* different lines to be able to apply it. You need one line that states the identity relation between two names and another line that has one of those two names coupled with a predicate.

1. Ba
2. a=d      | Bd |
3. Bd       1, 2 IR

**Identity Replacement (IR):**
a=b
Pa    | Pb |

This completes our rules. A table of all the Proof Rules is presented in the Unit 3 Review.

Note: Remember, all rules may be used. In working with quantifiers, you may still use the original implication and equivalence rules, as well as Conditional Proof and *Reductio* Proof, where they apply.

**STRATEGY TIPS**

Just as we did in the previous chapter, we'll conclude this chapter by adding three more strategy tips to the eight we provided in the last chapter to help you when you approach a proof.

**EE before UC**: As with trees, you'll need almost always to apply EE before UC. If you don't you're almost certain to get into trouble.

**What kind of statement is the conclusion?**: If you have a quantified statement as your conclusion, look at what type it is. If it's a universal statement you may have to use UC as your final move. If it's an existential statement, you may have to use EC.

**Try negating the conclusion**: Consider negating the conclusion and using RAA. You know that the proof is valid (we're only giving you valid proofs), so you should be able to eventually generate a contradiction.

# Exercise 13c

Construct proofs for the following arguments.

1. (x)(Nx → Dxx), (∃x)(Ex & Nx)   (∃x)(Ex & Dxx)
2. (x)(Ex → Dxt), (x)(Dxt → ~Ox)   (x)(Ex → ~Ox)
3. (∃x)Px, (x)(y)(Lxy ↔ Lxx), (x)(Px → ~Lmx)   ~Lmm
4. (x)[Px → ~(y)(Py → Lxy)]   ~(∃x)[Px & (y)(Py → Lxy)]
5. (x)(Px → ~(∃y)Mxy)   ~(∃x)(Px & Mxb)
6. (x)(Kbx → Gxc), (∃x)Gxc → (∃y)Gcy   (∃x)Kbx → (∃y)Gcy
7. (∃y)(x)(Lxy → Mxy)   (∃x)(y)Lxy → (∃y)(∃x)Mxy
8. Na & ~Nb   ~a=b
9. c=d → e=g, d=c, Fg   Fe
10. (∃x)x=a → ~b=b   ~a=b
11. (z)(Cz → Dz), ~Dg & Ca   ~g=a
12. (x)(Hx → Jx), (x)(Kx → Lx), Hd & Kc, c=d   Jc & Ld
13. (x)(Gx → x=d), (∃x)(Fx & Gx)   Fd
14. (x)(y)[Cyx → ~(z)Dz], (x)n=x, (∃x)[Ax & (y)(Ay → Cxy)]   ~Dn

15. (x)x=a   ⟨(x)(y)x=y⟩
16. (x)[(∃y)Kxy → (∃z)Kzx], Kbb   ⟨(∃z)Kzb⟩
17. (x)(y)y=x, (x)Mxx   ⟨Mab⟩
18. (x)[Mx → (x=a v x=b)], (∃x)(Mx & Nx)   ⟨Ma v Mb⟩
19. (x)(~Gx → ~x=d)   ⟨Gd⟩
20. (∃x)(y)Cyx   ⟨(x)(∃y)Cxy⟩

# Unit Three Review

## COMMON ARGUMENT INDICATORS

| Premise Indicators ||
| --- | --- |
| As | Firstly (secondly, thirdly) |
| Since | Seeing that |
| For | For the reasons that |
| Because | In view of the fact that |
| As shown by | On the correct supposition that |
| As indicated by | Assuming that |
| Follows from | May be inferred from |
| Being that | May be deduced from |
| Being as | May be derived from |
| Inasmuch as | Whereas |
| In the first (2nd, 3rd) place | |

| Conclusion Indicators ||
| --- | --- |
| Consequently | Implies that |
| Therefore | Entails that |
| Hence | Accordingly |
| So | I conclude that |
| Thus | You see that |
| Then | Leads me to believe that |
| Which shows that | Bears out the point that |
| Points to the conclusion that | Demonstrates that |
| Allows us to infer that | It follows that |
| Proves that | One then sees that |

NOTE: Seeing these words in a discourse doesn't guarantee that you're dealing with an argument; many of these words are used in different contexts. These words do, however, appear in many arguments. Also, these lists are not exhaustive; other words may be used to indicate premises or conclusions.

## VALIDITY

Df: Property of an argument where the conclusion cannot be false if the premises are true.

Abbreviated Table: Impossible to assign truth-values that make all the premises true and the conclusion false.

Tree: Tree with all true premises and the denial of the conclusion yields all closed branches.

**Table Showing Validity**

| A | B | A → B | A | B |
|---|---|-------|---|---|
| F | F | F T F | F | F |

No assignment of values makes both premises true and conclusion false.

**Table Showing Invalidity**

| A | B | A → B | ~A | ~B |
|---|---|-------|----|----|
| F | T | F T T | T F | F T |

At least one assignment of truth values makes premises true and conclusion false.

**Tree Showing Validity**

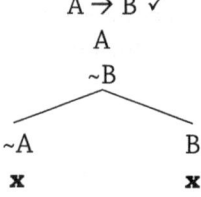

**Tree Showing Invalidity**

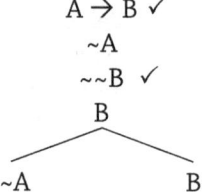

## PROOFS TO ESTABLISH VALIDITY

| PROOF RATIONALE | | PROOF SET UP |
|---|---|---|
| From a given set of premises, if a conclusion can be reached through a series of VALID steps, the argument as a whole is VALID. | 1. S <br> 2. S → U <br> 3. U → A <br> 4. A → K  $\boxed{K}$ | Number lines, <br> List all premises, <br> "Box" conclusion at end of the line of last premise. |

## PROOF RULES

*Basic Implication Rules*

| Modus Ponens (MP) | Modus Tollens (MT) |
|---|---|
| A → B <br> A    $\boxed{B}$ | A → B <br> ~B    $\boxed{\sim A}$ |
| **Hypothetical Syllogism (HS)** | **Disjunctive Syllogism (DS)** |
| A → B <br> B → C   $\boxed{A \to C}$ | A v B <br> ~A    $\boxed{B}$ |
| **Dilemma (DI)** | **Simplification (SIMP)** |
| A v B <br> A → C <br> B → D   $\boxed{C \vee D}$ | A & B   $\boxed{A}$ <br><br> OR:   A & B   $\boxed{B}$ |
| **Conjunction (CONJ)** | **Addition (ADD)** |
| A <br> B   $\boxed{A \& B}$ <br><br> OR:   A <br> B   $\boxed{B \& A}$ | A   $\boxed{A \vee B}$ <br><br> OR   A   $\boxed{B \vee A}$ |

*Basic Equivalence Rules*

| Commutation (COM) | Association (AS) |
|---|---|
| A v B :: B v A <br> A & B :: B & A | (A v B) v C :: A v (B v C) <br> (A & B) & C :: A & (B & C) |
| **Contraposition (CONT)** | **Double Negation (DN)** |
| A → B :: ~B → ~A | ~~A :: A |
| **Biconditional Breakdown (BCB)** | **Material Implication (MI)** |
| A ↔ B :: (A → B) & (B → A) | ~A v B :: A → B |
| **Duplication (DUP)** | **DeMorgan Crank (DeM/CR)** |
| A :: A v A <br> A :: A & A | ~(A v B) :: ~A & ~B <br> ~(A & B) :: ~A v ~B |

*Quantifier and Identity Rules*

| Universal Elimination (UE) | Existential Creation (EC) |
|---|---|
| (x)Ax    Aa | Aa    (∃x)Ax |
| **Quantifier Negation (QN)** | **Identity Introduction (II)** |
| ~(x)Px :: (∃x)~Px <br> ~(∃x) Px :: (x)~Px | n=n |
| **Existential Elimination (EE)** | **Universal Creation (UC)** |
| (∃x)Ax    Aa <br> 'a' must be new to proof, and must not appear in its conclusion. | Aa    (x)Ax <br> 'a' being replaced must have been arrived at by Universal Elimination. |
| **Symmetry (SYM)** | **Identity Replacement (IR)** |
| a=b :: b=a <br> ~a=b :: ~b=a | a=b <br> Aa    Ab |

# Unit Three: Answers to Selected Problems

## CHAPTER ELEVEN

### Exercise 11a

2. If Clark Kent is identical to Superman, then he's a superhero. Clark Kent is the secret identity of Superman. So, Clark Kent is a superhero.

$$c=s \rightarrow Hc, \quad c=s \quad \boxed{Hc}$$

$$c=s \rightarrow Hc \;\checkmark$$
$$c=s \quad\quad \boxed{Hc}$$
$$\sim Hc$$

```
        ~c=s        Hc
         x           x
```

**VALID: Tree with premises and denial of conclusion yields all closed branches, showing it is impossible for the premises to be true and the conclusion false.**

5. Batman is a great detective, even though Superman is stronger than everyone. So, it's still appropriate to conclude that Batman is a great detective.

$$Db \;\&\; (x)(\sim x=s \rightarrow Ssx) \quad \boxed{Db}$$

$$Db \;\&\; (x)(\sim x=s \rightarrow Ssx) \;\checkmark \quad \boxed{Db}$$
$$\sim Db$$
$$Db$$
$$(x)(\sim x=s \rightarrow Ssx)$$
$$x$$

**VALID: Tree with premises and denial of conclusion yields all closed branches, showing it is impossible for the premises to be true and the conclusion false.**

7. Either Batman is a great detective or he has superhuman strength. He doesn't have superhuman strength. So, he's a great detective.
**Db v (x)(~x=b → Sbx), ~(x)(~x=b → Sbx)** ☐ **Db**

        Db v (x)(~x=b → Sbx)  ✓
          ~(x)(~x=b → Sbx)       ☐ Db
              ~Db
            /    \
        Db     (x)(~x=b → Sbx)
         x               x

**VALID: Tree with premises and denial of conclusion yields all closed branches, showing it is impossible for the premises to be true and the conclusion false.**

8. If Clark Kent is identical to Superman, then Clark Kent is vulnerable to kryptonite. If Clark Kent is vulnerable to kryptonite, then he has a fatal weakness. So, Clark Kent has a fatal weakness.
**c=s → Vc, Vc → Wc** ✓ ☐ **Wc**

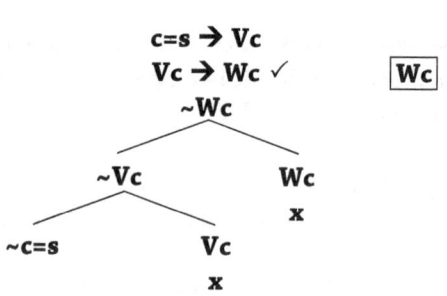

**INVALID: Tree with premises and denial of conclusion yields an open branch, showing it is possible for the premises to be true and the conclusion false.**

## Exercise 11b

1. If at least one thing is fun, then there's a least one thing that isn't dull. If there is at least one thing that isn't dull, then a person doesn't have to be bored. Hence, if at least one thing is fun, then people don't have to be bored.

$$(\exists x)Ux \rightarrow (\exists x){\sim}Dx, \; (\exists x){\sim}Dx \rightarrow (x)(Px \rightarrow {\sim}Bx)$$
$$\boxed{(\exists x)Ux \rightarrow (x)(Px \rightarrow {\sim}Bx)}$$

### HYPOTHETICAL SYLLOGISM (HS)

$(\exists x)Ux \rightarrow (\exists x){\sim}Dx$ ✓ $\boxed{(\exists x)Ux \rightarrow (x)(Px \rightarrow {\sim}Bx)}$
$(\exists x){\sim}Dx \rightarrow (x)(Px \rightarrow {\sim}Bx)$ ✓
${\sim}[(\exists x)Ux \rightarrow (x)(Px \rightarrow {\sim}Bx)]$ ✓
$(\exists x)Ux$
${\sim}(x)(Px \rightarrow {\sim}Bx)$

```
         ~(∃x)Ux              (∃x)~Dx
            x           ┌────────┴────────┐
                     ~(∃x)~Dx      (x)(Px → ~Bx)
                        x                x
```

**VALID**

4. Either everyone is fun, or someone is dull. It's not the case that everyone is fun. Consequently, someone is dull.

$$(x)(Px \rightarrow Ux) \lor (\exists x)Dx, \; {\sim}(x)(Px \rightarrow Ux) \quad \boxed{(\exists x)Dx}$$

### DISJUNCTIVE SYLLOGISM (DS)

$(x)(Px \rightarrow Ux) \lor (\exists x)Dx$ ✓ $\boxed{(\exists x)Dx}$
${\sim}(x)(Px \rightarrow Ux)$
${\sim}(\exists x)Dx$

```
      (x)(Px → Ux)      (∃x)Dx
           x               x
```

**VALID**

8. Either everyone is free or no one is free. If everyone is free, then we don't need to fight. If no one is free, then everyone is confused. Thus, either we don't need to fight or everyone is confused.

(x)(Px → Fx) v (x)(Px → ~Fx), (x)(Px → Fx) → (x)(Px → ~Ix), (x)(Px → ~Fx) → (x)(Px → Cx)   |(x)(Px → ~Ix) v (x)(Px → Cx)|

**DILEMMA**

(x)(Px → Fx) v (x)(Px → ~Fx)   ✓   |(x)(Px → ~Ix) v (x)(Px →|
(x)(Px → Fx) → (x)(Px → ~Ix)   ✓
(x)(Px → ~Fx) → (x)(Px → Cx)   ✓
~[(x)(Px → ~Ix) v (x)(Px → Cx)]   ✓
~(x)(Px → ~Ix)
~(x)(Px → Cx)

```
              ~(x)(Px → Fx)          (x)(Px → ~Ix)
                                           x
  (x)(Px → Fx)     (x)(Px → ~Fx)
        x
              ~(x)(Px → ~Fx)     (x)(Px → Cx)
                    x                  x
```

**VALID**

9. Someone wants to be friends with John, which we can see for the following reasons. If someone is fun, then they'll want to be friends with John. There has to be someone out there who is fun.
(x)[(Px & Ux) → Fxj], (∃x)(Px & Ux)   $\boxed{(\exists x)Fxj}$

**MODUS PONENS-ish (MP)**

$$
\begin{array}{c}
(x)[(Px \& Ux) \rightarrow Fxj] \quad * \quad \boxed{(\exists x)Fxj} \\
(\exists x)(Px \& Ux) \quad \checkmark \\
\sim(\exists x)Fxj \quad \checkmark \\
(x)\sim Fxj \quad * \\
Pr \& Ur \quad \checkmark \\
Pr \\
Ur \\
\sim Frj \\
(Pr \& Ur) \rightarrow Frj
\end{array}
$$

```
          ~(Pr & Ur)        Frj
                             x
        ~Pr        ~Ur
         x          x
```

**VALID**

## Exercise 11c

3. A → (B v ~D), A → C   $\boxed{B \rightarrow C}$

| A | B | C | D | A → (B v ~D) | A → C | B → C |
|---|---|---|---|---|---|---|
| F | T | F | F | F T  T T T F | F T F | T F F |

**INVALID:** *Possible* to construct a row with *true premises* and a *false conclusion*.

5. A v B, A → C, B → D   $\boxed{C \vee D}$

| A | B | C | D | A v B | A → C | B → D | C v D |
|---|---|---|---|---|---|---|---|
| F | F | F | F | F F F | F T F | F T F | F F F |

**VALID:** *Impossible* to construct a row with true premises and a false conclusion. In order to make premises 2 or 3 true, premise 1 is false. If you made premise 1 true, then either premise 2 or 3 would be false.

6. ~A → B   $\boxed{A \& B}$

| A | B | ~A → B | A & B |
|---|---|--------|-------|
| F | F | T F F F | F F F |
| F | T | T F T T | F F T |

**INVALID:** *Possible* to construct a row (ROW 2) with *true premises* and a *false conclusion*.

7. B → (A & B), (C v A) & B   $\boxed{A \leftrightarrow B}$

| A | B | C | B → (A & B) | (C v A) & B | A ↔ B |
|---|---|---|-------------|-------------|-------|
| T | F |   | F T T F F   | T   F F     | T F F |
| F | T |   | T F F F     |             | F F T |

**VALID:** *Impossible* to construct a row with true premises and a false conclusion. No matter what truth value we'd assign for C, the truth assignments for A and B needed to make the conclusion false would always yield a false premise.

9. C v B, A & ~B   $\boxed{C}$

| C | B | A | C v B | A & ~B | C |
|---|---|---|-------|--------|---|
| F | T |   | F T T | F F    | F |

**VALID:** *Impossible* to construct a row with true premises and a false conclusion. We have to make B true in order to make premise 1 true, but that makes premise 2 false (because ~B is false).

13. K ↔ M, M → L, L v ~M   $\boxed{M \vee L}$

| K | M | L | K ↔ M | M → L | L v ~M | M v L |
|---|---|---|-------|-------|--------|-------|
| F | F | F | F T F | F T F | F T T F | F F F |

**INVALID:** *Possible* to construct a row with *true premises* and a *false conclusion*.

16. P & ~Q, Q v (P → M)    $\boxed{\text{M v Q}}$

| P | Q | M | P & ~Q | Q v (P → M) | M v Q |
|---|---|---|--------|-------------|-------|
| T | F | F | T T T F | F F T F F | F F F |

**VALID:** *Impossible* to construct a row with true premises and a false conclusion. Truth value assignments that make conclusion false and first premise true make the 2nd premise false.

## CHAPTER TWELVE

**Exercise 12a**

1. Ma, Ma → (x)(Px → Kx), (x)(Px → Kx) → (∃x)Lx, (∃x)Lx → Ta    $\boxed{\text{Ta}}$
   1. Ma
   2. Ma → (x)(Px → Kx)
   3. (x)(Px → Kx) → (∃x)Lx
   4. (∃x)Lx → Ta                           $\boxed{\text{Ta}}$
   5. Ma → (∃x)Lx              2, 3 HS
   6. Ma → Ta                  4, 5 HS
   7. Ta                       1, 6 MP

7. Np → (Fo v ~Os), ~~Os & Np    $\boxed{\text{Fo v ~La}}$
   1. Np → (Fo v ~Os)
   2. ~~Os & Np                              $\boxed{\text{Fo v ~La}}$
   3. Np                       2, SIMP
   4. Fo v ~Os                 1, 3 MP
   5. ~~Os                     2, SIMP
   6. Fo                       4, 5 DS
   7. Fo v ~La                 6, ADD

8. [(∃x)Bx & (y)Zy] → Qa, (∃x)Bx, Qa → (Ci & Ce), (y)Zy    $\boxed{\text{(Ci & Ce)}}$
   1. [(∃x)Bx & (y)Zy] → Qa
   2. (∃x)Bx
   3. Qa → (Ci & Ce)
   4. (y)Zy                                  $\boxed{\text{(Ci & Ce)}}$
   5. [(∃x)Bx & (y)Zy] → (Ci & Ce)    1, 3 HS
   6. (∃x)Bx & (y)Zy           2, 4 CONJ
   7. Ci & Ce                  5, 6 MP

9. ~Nr v Gr, ~Nr → Et, Gr → ~Hs, (Et v ~Hs) → (Nq → Ga), ~Ga   ~Nq
   1. ~Nr v Gr
   2. ~Nr → Et
   3. Gr → ~Hs
   4. (Et v ~Hs) → (Nq → Ga)
   5. ~Ga                                          ~Nq
   6. Et v ~Hs                       1, 2, 3 DI
   7. Nq → Ga                        4, 6 MP
   8. ~Nq                            5, 7 MT

10. (~Vc & ~Ke) → (Be → Ke), ~Vc, Ke → Vc   ~Be
    1. (~Vc & ~Ke) → (Be → Ke)
    2. ~Vc
    3. Ke → Vc                                     ~Be
    4. ~Ke                           2, 3 MT
    5. ~Vc & ~Ke                     2, 4 CONJ
    6. Be → Ke                       1, 5 MP
    7. ~Be                           4, 6 MT

11. ~Dq & ~~Ba, (~Dq & Rb) → ~Ar, ~Ba v Rb   ~Ar
    1. ~Dq & ~~Ba
    2. (~Dq & Rb) → ~Ar
    3. ~Ba v Rb                                    ~Ar
    4. ~~Ba                          1, SIMP
    5. Rb                            3, 4 DS
    6. ~Dq                           1, SIMP
    7. ~Dq & Rb                      5, 6 CONJ
    8. ~Ar                           2, 7 MP

15. (Ad v Gd) → (z)Mza, (z)Mza → (Bf → Fd), Ad & Bf   Fd
    1. (Ad v Gd) → (z)Mza
    2. (z)Mza → (Bf → Fd)
    3. Ad & Bf                                     Fd
    4. Ad                            3, SIMP
    5. Bf                            3, SIMP
    6. Ad v Gd                       4, ADD
    7. (z)Mza                        1, 6 MP
    8. Bf → Fd                       2, 7 MP
    9. Fd                            5, 8 MP

18. (x)Px, (x)Px → Al, Al → (Eb v Fb), ~Eb   |Fb|
    1. (x)Px
    2. (x)Px → Al
    3. Al → (Eb v Fb)
    4. ~Eb                                    |Fb|
    5. (x)Px → (Eb v Fb)        2, 3 HS
    6. Eb v Fb                  1, 5 MP
    7. Fb                       4, 6 DS

19. Pa v Qa, (Qa & ~Rg) → Sb, Rg → Pa, ~Pa   |Sb|
    1. Pa v Qa
    2. (Qa & ~Rg) → Sb
    3. Rg → Pa
    4. ~Pa                                    |Sb|
    5. Qa                       1, 4 DS
    6. ~Rg                      3, 4 MT
    7. Qa & ~Rg                 5, 6 CONJ
    8. Sb                       2, 7 MP


**Exercise 12b**

2. (Ad v Bq) v Ct, ~Ad      |Ct v Bq|
    1. (Ad v Bq) v Ct
    2. ~Ad                                    |Ct v Bq|
    3. Ad v (Bq v Ct)           1, AS
    4. Bq v Ct                  2, 3 DS
    5. Ct v Bq                  4, COM

4. Bd → Ec, ~Fe v Gm, (Bd & Cm) & Da, (Da & Cm) → Fe   |Ec & Gm|
    1. Bd → Ec
    2. ~Fe v Gm
    3. (Bd & Cm) & Da
    4. (Da & Cm) → Fe                         |Ec & Gm|
    5. Bd & (Cm & Da)           3, AS
    6. Bd                       5, SIMP
    7. Ec                       1, 6 MP
    8. Cm & Da                  5, SIMP
    9. Da & Cm                  8, COM
   10. Fe                       4, 9 MP

|  |  |
|---|---|
| 11. ~~Fe | 10, DN |
| 12. Gm | 2, 11 DS |
| 13. Ec & Gm | 7, 12 CONJ |

7. [~(Sd & Tb) v ~~Km] & (Tb & Sd)    ☐ Km ☐
    1. [~(Sd & Tb) v ~~Km] & (Tb & Sd)    ☐ Km ☐
    2. ~(Sd & Tb) v ~~Km      1, SIMP
    3. Tb & Sd      1, SIMP
    4. Sd & Tb      3, COM
    5. ~~(Sd & Tb)      4, DN
    6. ~~Km      2, 5 DS
    7. Km      6, DN

8. (Ab & Ca) → Ba, (~Ba v ~Ab) v Dab, Ca & Ab    ☐ ~~(Dab v Pab) ☐
    1. (Ab & Ca) → Ba
    2. (~Ba v ~Ab) v Dab
    3. Ca & Ab    ☐ ~~(Dab v Pab) ☐
    4. Ab & Ca      3, COM
    5. Ba      1, 4 MP
    6. ~Ba v (~Ab v Dab)      2, AS
    7. ~~Ba      5, DN
    8. ~Ab v Dab      6, 7 DS
    9. Ab      3, SIMP
    10. ~~Ab      9, DN
    11. Dab      8, 10 DS
    12. Dab v Pab      11, ADD
    13. ~~(Dab v Pab)      12, DN

**Exercise 12c**

1. ~(Ca & Da), (~Ca → Sb), ~Da → Td    ☐ Sb v Td ☐
    1. ~(Ca & Da)
    2. ~Ca → Sb
    3. ~Da → Td    ☐ Sb v Td ☐
    4. ~Ca v ~Da      1, DM
    5. Sb v Td      2, 3, 4 DI

3. Pa ⟷ ~(x)Qx, ~(Pa v St)   / (x)Qx
    1. Pa ⟷ ~(x)Qx
    2. ~(Pa v St)
    3. ~Pa & ~St              2, DM
    4. ~(x)Qx → Pa            1, BCB
    5. ~Pa                    3, SIMP
    6. ~~(x)Qx                4, 5 MT
    7. (x)Qx                  6, DN

8. ~Ia ⟷ Ag, (Fn & Hj) v ~Ia, ~(Ag v Bn)   / Fn & ~Bn
    1. ~Ia ⟷ Ag
    2. (Fn & Hj) v ~Ia
    3. ~(Ag v Bn)                           / Fn & ~Bn
    4. ~Ag & ~Bn              3, DM
    5. ~Ag                    4, SIMP
    6. ~Bn                    4, SIMP
    7. ~(Fn & Hj) → ~Ia       2, MI
    8. ~Ia → Ag               1, BCB
    9. ~(Fn & Hj) → Ag        7, 8 HS
    10. ~~(Fn & Hj)           5, 9 MT
    11. Fn & Hj               10, DN
    12. Fn                    11, SIMP
    13. Fn & ~Bn              6, 12 CONJ

9. (Pd v Qb), Qb → St, Pd → (Ur v Ve), ~(St v Ve)   / Ur
    1. Pd v Qb
    2. Qb → St
    3. Pd → (Ur v Ve)
    4. ~(St v Ve)                           / Ur
    5. (Ur v Ve) v St         1, 2, 3 DI
    6. Ur v (Ve v St)         5, AS
    7. Ur v (St v Ve)         6, COM
    8. Ur                     4, 7 DS

11. (Pa & Ga) → Re, (Re & Sd) → Ta, Pa & Sd, Ga v Re   / Re v Ta
    1. (Pa & Ga) → Re
    2. (Re & Sd) → Ta
    3. Pa & Sd
    4. Ga v Re                              / Re v Ta
    5. ~(Pa & Ga) v Re        1, MI

| | | |
|---|---|---|
| 6. | (~Pa v ~Ga) v Re | 5, DM |
| 7. | ~Pa v (~Ga v Re) | 6, AS |
| 8. | Pa | 3, SIMP |
| 9. | ~~Pa | 8, DN |
| 10. | ~Ga v Re | 7, 9 DS |
| 11. | Ga → Re | 10, MI |
| 12. | ~(Re & Sd) v Ta | 2, MI |
| 13. | (~Re v ~Sd) v Ta | 12, DM |
| 14. | (~Sd v ~Re) v Ta | 13, COM |
| 15. | ~Sd v (~Re v Ta) | 14, AS |
| 16. | Sd | 3, SIMP |
| 17. | ~~Sd | 16, DN |
| 18. | ~Re v Ta | 15, 17 DS |
| 19. | Re → Ta | 18, MI |
| 20. | Re v Ta | 4, 11, 19 DI |

12. ~(Ca v Da), Da ↔ (Ep v Fm), ~Ak → (Ca v Fm)   ☐ Ak

| | | |
|---|---|---|
| 1. | ~(Ca v Da) | |
| 2. | Da ↔ (Ep v Fm) | |
| 3. | ~Ak → (Ca v Fm) | ☐ Ak |
| 4. | ~Ca & ~Da | 1, DM |
| 5. | (Ep v Fm) → Da | 2, BCB |
| 6. | ~Da | 4, SIMP |
| 7. | ~(Ep v Fm) | 5, 6 MT |
| 8. | ~Ep & ~Fm | 7, DM |
| 9. | ~Ca | 4, SIMP |
| 10. | ~Fm | 8, SIMP |
| 11. | ~Ca & ~Fm | 9, 10 CONJ |
| 12. | ~(Ca v Fm) | 11, DM |
| 13. | ~~Ak | 3, 12 MT |
| 14. | Ak | 13, DN |

14. ~(Ja & Tf), ~(~Tf v Ea)   ☐ ~(Ja v Ea)

| | | |
|---|---|---|
| 1. | ~(Ja & Tf) | |
| 2. | ~(~Tf v Ea) | ☐ ~(Ja v Ea) |
| 3. | ~Ja v ~Tf | 1, DM |
| 4. | ~~Tf & ~Ea | 2, DM |
| 5. | ~~Tf | 4, SIMP |
| 6. | ~Ja | 3, 5 DS |
| 7. | ~Ea | 4, SIMP |
| 8. | ~Ja & ~Ea | 6, 7 CONJ |
| 9. | ~(Ja v Ea) | 8, DM |

**Exercise 12d**

6. Pa → Qm, ~Pa → Ja, ~Qm → ~Ja      ⟦Qm⟧
   1. Pa → Qm
   2. ~Pa → Ja
   3. ~Qm → ~Ja                                      ⟦Qm⟧
   | 4. ~Qm | PA |
   | 5. ~Ja | 3, 4 MP |
   | 6. ~~Pa | 2, 5 MT |
   | 7. Pa | 6, DN |
   | 8. Qm | 1, 7 MP |
   | 9. Qm & ~Qm | 4, 8 CONJ |
   10. Qm                                            4 – 9 RAA

8. [(∃x)Mx → (∃x)Ky] → (∃x)Mx       ⟦(∃x)Mx⟧
   1. [(∃x)Mx → (∃y)Ky] → (∃x)Mx                    ⟦(∃x)Mx⟧
   | 2. ~(∃x)Mx | PA |
   | 3. ~[(∃x)Mx → (∃y)Ky] | 1, 2 MT |
   | 4. ~[~(∃x)Mx v (∃y)Ky] | 3, MI |
   | 5. (∃x)Mx & ~(∃y)Ky | 4, DM |
   | 6. (∃x)Mx | 5, SIMP |
   | 7. (∃x)Mx & ~(∃x)Mx | 2, 6 CONJ |
   8. (∃x)Mx                                        2 – 7 RAA

9. ~Ac & ~Ba          ⟦Ac ↔ Ba⟧
   1. ~Ac & ~Ba                                     ⟦Ac ↔ Ba⟧
   | 2. Ac | PA |
   | 3. Ac v Ba | 2, ADD |
   | 4. ~Ac | 1, SIMP |
   | 5. Ba | 3, 4 DS |
   6. Ac → Ba                                       2 – 5 CP
   | 7. Ba | PA |
   | 8. Ba v Ac | 6, ADD |
   | 9. ~Ba | 1, SIMP |
   | 10. Ac | 7, 8 DS |
   11. Ba → Ac                                      7 – 10 CP
   12. (Ac → Ba) & (Ba → Ac)                        6, 11 CONJ
   13. Ac ↔ Ba                                      12, BCB

10. ~Pr → (Ra & Sa), ~Qr → (Ra & Tr), ~(Sa v Tr)        /  Pr & Qr
    1. ~Pr → (Ra & Sa)
    2. ~Qr → (Ra & Tr)
    3. ~(Sa v Tr)                                          / Pr & Qr
    4. ~Sa & ~Tr              3, DM
    5. ~Sa                    4, SIMP
    6. ~Tr                    4, SIMP
    7. ~Pr                    PA
    8. Ra & Sa                1, 7 MP
    9. Sa                     8, SIMP
    10. Sa & ~Sa              5, 9 CONJ
    11. Pr                    7 – 10 RAA
    12. ~Qr                   PA
    13. Ra & Tr               2, 12 MP
    14. Tr                    13, SIMP
    15. Tr & ~Tr              6, 14 CONJ
    16. Qr                    12 – 15 RAA
    17. Pr & Qr               11, 16 CONJ

13. (~Mo v ~Bg) → ~Cg        / Cg → Mo
    1. (~Mo v ~Bg) → ~Cg                                  / Cg → Mo
    2. Cg                     PA
    3. ~~Cg                   2, DN
    4. ~(~Mo v ~Bg)           1, 3 MT
    5. Mo & Bg                4, DM
    6. Mo                     5, SIMP
    7. Cg → Mo                2 – 6 CP

15. (Kc v Na) → ~Ce, Dh → (~Fab & ~Gj)  ⬚ (Kc v Dh) → ~(Ce & Fab)
    1. (Kc v Na) → ~Ce
    2. Dh → (~Fab & ~Gj)  ⬚ (Kc v Dh) → ~(Ce & Fab)

| 3. Kc v Dh | PA |
|---|---|
| 4. Ce & Fab | PA |
| 5. Ce | 4, SIMP |
| 6. ~~Ce | 5, DN |
| 7. ~(Kc v Na) | 1, 6 MT |
| 8. ~Kc & ~Na | 7, DM |
| 9. ~Kc | 8, SIMP |
| 10. Dh | 3, 9 DS |
| 11. ~Fab & ~Gj | 2, 10 MP |
| 12. ~Fab | 11, SIMP |
| 13. Fab | 4, SIMP |
| 14. Fab & ~Fab | 12, 13 CONJ |
| 15. ~(Ce & Fab) | 4 – 14 RAA |

    16. (Kc v Dh) → ~(Ce & Fab)    3 – 15 CP

17. (Ak v Ql) → ~(Gd & Dj), (Ak & Dj), ~Gd → ~(Ca & Dj)  ⬚ ~(Ak & Ca)
    1. (Ak v Ql) → ~(Gd & Dj)
    2. Ak & Dj
    3. ~Gd → ~(Ca & Dj)  ⬚ ~(Ak & Ca)

| 4. Ak & Ca | PA |
|---|---|
| 5. Ak | 4, SIMP |
| 6. Ca | 4, SIMP |
| 7. Ak v Ql | 5, ADD |
| 8. ~(Gd & Dj) | 1, 7 MP |
| 9. ~Gd v ~Dj | 8, DM |
| 10. Dj | 2, SIMP |
| 11. ~~Dj | 10, DN |
| 12. ~Gd | 9, 11 DS |
| 13. ~(Ca & Dj) | 3, 12 MP |
| 14. ~Ca v ~Dj | 13, DM |
| 15. ~Ca | 11, 14 DS |
| 16. Ca & ~Ca | 6, 15 CONJ |
| 17. ~(Ak & Ca) | 4 – 16 RAA |

19. Abg ⟷ ~(Bh v Cs), Bh ⟷ (Ds & ~Epo), ~(Epo & Abg)   | Abg → ~Ds |
   1. Abg ⟷ ~(Bh v Cs)
   2. Bh ⟷ (Ds & ~Epo)
   3. ~(Epo & Abg)                                              | Abg → ~Ds |
   4. Abg                                     PA
   5. Abg → ~(Bh v Cs)                        1, BCB
   6. ~(Bh v Cs)                              4, 5 MP
   7. ~Bh & ~Cs                               6, DM
   8. ~Bh                                     7, SIMP
   9. (Ds & ~Epo) → Bh                        2, BCB
  10. ~(Ds & ~Epo)                                 8, 9 MT
  11. ~Ds v Epo                                    10, DM
  12. ~Epo v ~Abg                                  3, DM
  13. ~~Abg                                        4, DN
  14. ~Epo                                         12, 13 DS
  15. ~Ds                                          11, 14 DS
  16. Abg → ~Ds                                    4 – 15 CP

## CHAPTER THIRTEEN

**Exercise 13a**

  1. (x)(y)(z)(Mxy → Myz), Mlj   | Mjl |
    1. (x)(y)(z)(Mxy → Myz)
    2. Mlj                                              | Mjl |
    3. (y)(z)(Mly → Myz)         1, UE
    4. (z)(Mlj → Mjz)            3, UE
    5. Mlj → Mjl                 4, UE
    6. Mjl                       2, 5 MP

  4. (y)(Bay v Bya)   | Baa |
    1. (y)(Bay v Bya)                                   | Baa |
    2. Baa v Baa                 1, UE
    3. Baa                       2, DUP

7. (x)(Tx → (y)[Vy → (∃z)(Pz & Syz)]), Tb & Ve    (∃z)(Pz & Sez)
   1. (x)(Tx → (y)[Vy → (∃z)(Pz & Syz)])
   2. Tb & Ve                                      (∃z)(Pz & Sez)
   3. Tb → (y)[Vy → (∃z)(Pz & Syz)]       1, UE
   4. Tb                                   2, SIMP
   5. (y)[Vy → (∃z)(Pz & Syz)]            3, 4 MP
   6. Ve → (∃z)(Pz & Sez)                 5, UE
   7. Ve                                   2, SIMP
   8. (∃z)(Pz & Sez)                       6, 7 MP

8. ~(x)(y)Pxy, ~Dab → (x)(y)Pxy    (∃x)(∃y)Dxy
   1. ~(x)(y)Pxy
   2. ~Dab → (x)(y)Pxy                              (∃x)(∃y)Dxy
   3. ~~Dab                                1, 2 MT
   4. Dab                                  3, DN
   5. (∃y)Day                              4, EC
   6. (∃x)(∃y)Dxy                          5, EC

9. (y)(Oyd → Cy), Ocd & Pcd, (z)(~Pzd v Lz)    (∃x)(Lx & Cx)
   1. (y)(Oyd → Cy)
   2. Ocd & Pcd
   3. (z)(~Pzd v Lz)                                (∃x)(Lx & Cx)
   4. Ocd → Cc                             1, UE
   5. Ocd                                  2, SIMP
   6. Cc                                   4, 5 MP
   7. ~Pcd v Lc                            3, UE
   8. Pcd                                  2, SIMP
   9. ~~Pcd                                8, DN
   10. Lc                                  7, 9 DS
   11. Lc & Cc                             6, 10 CONJ
   12. (∃x)(Lx & Cx)                       11, EC

**Exercise 13b**

1. (x)(Ax & Bx)    (x)Ax & (x)Bx
   1. (x)(Ax & Bx)                                  (x)Ax & (x)Bx
   2. Ar & Br                              1, UE
   3. Ar                                   2, SIMP
   4. Br                                   2, SIMP
   5. (x)Ax                                3, UC
   6. (x)Bx                                4, UC
   7. (x)Ax & (x)Bx                        5, 6 CONJ

2. ~(x)~Ax, (∃x)Ax → (x)Bx, (x)[(Bx v Cx) → Dx]   ☐(x)Dx☐
   1. ~(x)~Ax
   2. (∃x)Ax → (x)Bx
   3. (x)[(Bx v Cx) → Dx]                          ☐(x)Dx☐
   4. (∃x)~~Ax                    1, QN
   5. (∃x)Ax                      4, DN
   6. (x)Bx                       2, 5 MP
   7. Br                          6, UE
   8. (Br v Cr) → Dr              3, UE
   9. Br v Cr                     7, ADD
   10. Dr                         8, 9 MP
   11. (x)Dx                      10, UC

8. ~(∃x)~Px, ~(∃y)Sy v ~(x)Px,   ☐~Sd☐
   1. ~(∃x)~Px
   2. ~(∃y)Sy v ~(x)Px                             ☐~Sd☐
   3. ~~(x)Px                     1, QN
   4. ~(∃y)Sy                     2, 3 DS
   5. (y)~Sy                      4, QN
   6. ~Sd                         5, UE

10. (∃x)[Kx & (Ax & Sx)], (x)[(Ax v Sx) → ~Wx]   ☐(∃x)(Kx & ~Wx)☐
    1. (∃x)[Kx & (Ax & Sx)]
    2. (x)[(Ax v Sx) → ~Wx]                       ☐(∃x)(Kx & ~Wx)☐
    3. Ka & (Aa & Sa)             1, EE
    4. (Aa v Sa) → ~Wa            2, UE
    5. Ka & (Sa & Aa)             3, COM
    6. (Ka & Sa) & Aa             5, AS
    7. Aa                         6, SIMP
    8. Aa v Sa                    7, ADD
    9. ~Wa                        4, 8 MP
    10. Ka                        3, SIMP
    11. Ka & ~Wa                  9, 10 CONJ
    12. (∃x)(Kx & ~Wx)            11, EE

**Exercise 13c**

1. (x)(Nx → Dxx), (∃x)(Ex & Nx)         (∃x)(Ex & Dxx)
   1. (x)(Nx → Dxx)
   2. (∃x)(Ex & Nx)                     (∃x)(Ex & Dxx)
   3. Er & Nr                            2, EE
   4. Nr                                 3, SIMP
   5. Nr → Drr                           1, UE
   6. Drr                                4, 5 MP
   7. Er                                 3, SIMP
   8. Er & Drr                           6, 7 CONJ
   9. (∃x)(Ex & Dxx)                     8, EC

3. (∃x)Px, (x)(y)(Lxy ↔ Lxx), (x)(Px → ~Lmx)     ~Lmm
   1. (∃x)Px
   2. (x)(y)(Lxy ↔ Lxx)
   3. (x)(Px → ~Lmx)                     ~Lmm
   | 4. Lmm                              PA          |
   | 5. Pr                               1, EE       |
   | 6. Pr → ~Lmr                        3, UE       |
   | 7. ~Lmr                             5, 6 MP     |
   | 8. (y)(Lmy ↔ Lmm)                   2, UE       |
   | 9. Lmr ↔ Lmm                        7, UE       |
   | 10. Lmm → Lmr                       8, BCB      |
   | 11. Lmr                             4, 10 MP    |
   | 12. Lmr & ~Lmr                      7, 11 CONJ  |
   13. ~Lmm                              4 – 12 RAA

5. (x)(Px → ~(∃y)Mxy)                    ~(∃x)(Px & Mxb)
   1. (x)(Px → ~(∃y)Mxy)                 ~(∃x)(Px & Mxb)
   | 2. (∃x)(Px & Mxb)                   PA          |
   | 3. Pa & Mab                         2, EE       |
   | 4. Pa                               3, SIMP     |
   | 5. Mab                              3, SIMP     |
   | 6. Pa → ~(∃y)May                    1, UE       |
   | 7. ~(∃y)May                         4, 6 MP     |
   | 8. (∃y)May                          5, EC       |
   | 9. (∃y)May & ~(∃y)May                7, 8 CONJ   |
   10. ~(∃x)(Px & Mxb)                   2 – 9 RAA

7. (∃y)(x)(Lxy → Mxy)     ⊢ (∃x)(y)Lxy → (∃y)(∃x)Mxy
     1. (∃y)(x)(Lxy → Mxy)     (∃x)(y)Lxy → (∃y)(∃x)Mxy
     2. (∃x)(y)Lxy                PA
     3. (y)Lay                   2, EE
     4. (x)(Lxb → Mxb)       1, EE
     5. Lab                     3, UE
     6. Lab → Mab           4, UE
     7. Mab                    5, 6 MP
     8. (∃x)Mxb               7, EC
     9. (∃y)(∃x)Mxy          8, EC
    10. (∃x)(y)Lxy → (∃y)(∃x)Mxy     2 – 9 CP

8. Na & ~Nb     ⊢ ~a=b
     1. Na & ~Nb             ~a=b
     2. a=b                     PA
     3. Na                      1, SIMP
     4. Nb                      2, 3 IR
     5. ~Nb                    1, SIMP
     6. Nb & ~Nb             4, 5 CONJ
     7. ~a=b                 2 – 6 RAA

9. c=d → e=g, d=c, Fg     ⊢ Fe
     1. c=d → e=g
     2. d=c
     3. Fg                      Fe
     4. c=d                     2, Sym
     5. e=g                     1, 4 MP
     6. Fe                      3, 5 IR

10. (∃x)x=a → ~b=b     ⊢ ~a=b
     1. (∃x)x=a → ~b=b       ~a=b
     2. b=b                     II
     3. ~~b=b                 2, DN
     4. ~(∃x)x=a             1, 3 MT
     5. (x)~x=a              4, QN
     6. ~b=a                   5, UE
     7. ~a=b                   6, IS

16. (x)[(∃y)Kxy → (∃z)Kzx], Kbb　　　　$\boxed{(\exists z)Kzb}$
   1. **(x)[(∃y)Kxy → (∃z)Kzx]**
   2. **Kbb**　　　　　　　　　　　　$\boxed{(\exists z)Kzb}$
   3. **(∃y)Kby → (∃z)Kzb**　　1, UE
   4. **(∃y)Kby**　　　　　　　　2, EC
   5. **(∃z)Kzb**　　　　　　　　3, 4 MP

# Postscript: Logic in Real Life

> **What's Up?**
> Argument-Analysis Tree
> Deductive arguments:
> Translation and analysis

We've now completed this introductory look at the logic of our language. We began with showing the logic of grammatically very simple declarative sentences (e.g., "Freddy focused."), and worked up to extremely complex compound statements (e.g., "There are exactly two New Jersey senators."), using our artificial language, LOLA. We've examined how truth tables and trees demonstrate the truth values of compounds, as well as what "flavor" of truth they are (L-true, L-false, or contingent). Finally we've examined important relations between and among statements, ending with the very important relationship of implication, most particularly when premises imply conclusions in arguments. An argument's validity depends on this implication relation, and we learned how tables, trees, and proofs can help us establish an argument's validity.

But, where do we go from here? Working with a language, any language, takes practice—as you've probably discovered during this course. And a person's fluency and skillful use of a newly learned language will generally fade unless it is used. Why is this a problem? Well for one thing, as we said in the first chapter, learning symbolic logic and its methods of analysis can help us sort out longer and more challenging discourse, such as we find in many college courses. Arguments, for example, can be page length, chapter length, and even book length, and often the structure of such arguments can be very complex.

Since this is an introductory course, we won't ask you to do a chapter or book-length analysis here! But we offer a method to begin an analysis of longer arguments, called Argument-Analysis Trees. Like truth-trees, we ask a question at every step along the way. But instead of asking, "Under what conditions would this statement be true," you answer a series of yes-or-no questions which, depending how you respond, will direct you farther along the tree path.

After we present the tree, we'll go through a fairly short argument as an example, and we hope you will take it from there and continue to upgrade your logical skills. We'll finish off by offering a few short arguments that you can practice on.

# ARGUMENT-ANALYSIS TREE

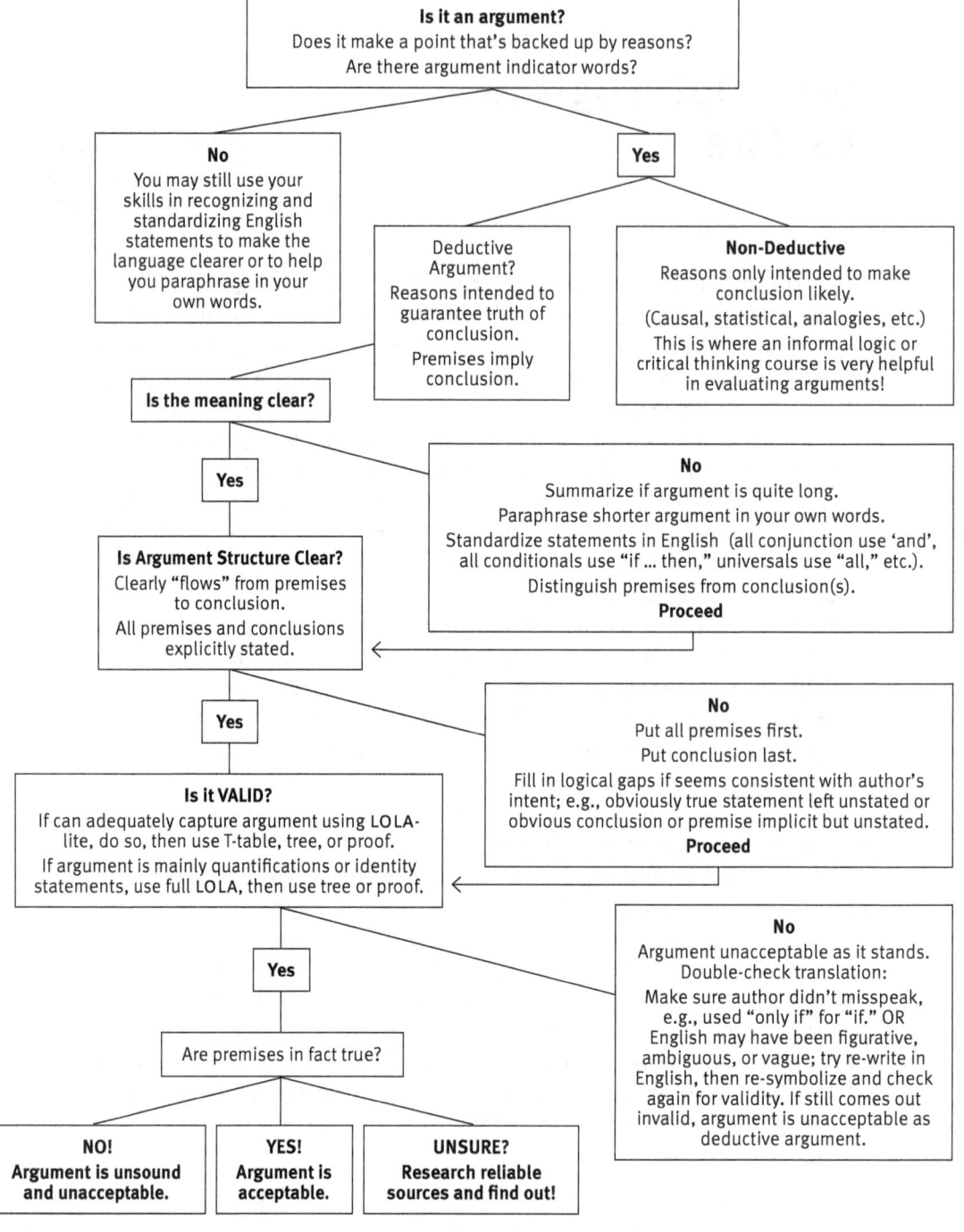

## ARGUMENT EXAMPLE

Jim Thompson is a true American original, and being such, is very difficult to classify. From his roots in Oklahoma and Texas he conveys the sensibility of the Western. But Thompson's West is not one of open expanse and unlimited possibility. His characters proceed as if on a crazy-quilt chessboard, moved by unseen players who care not whether they win or lose, and *all* the pieces are pawns. There are no rugged individuals here, merely isolated, broken remnants of humanity, blessed only with limited intelligence and varying degrees of guile.

Thompson's taut first-person narratives, crisp style, and femmes (and *hommes*) fatale, call to mind the classic hard-boiled detective story and *film noir*. But his *noir* landscape has no self-aware (albeit bitter) heroes, no coming to terms with alienation, no self-imposed codes in a world without ultimate meaning. God is not dead in Thompsonville—He is abandoned by the inhabitants, and He returns the favor. The *noir* city begs us to contrast it with the daylight city, populated by unbelievers who have not yet seen the dark. But Thompson's world does not beg such contrast; it compels us to contemplate this world as it is. There is no contrasting sunlight—a *monde gris*.

Jim Thompson has rightly been called a suspense writer, but the suspense is fueled by an unrelenting, grey, Western sun. We are held to the pace of the wandering, footsore tramp, the dusty, spent horse, the lumbering freight train. But the Thompsonville Choo-Choo is heading toward a precipice—we watch, we wait, we expect the plunge, the crash, the smoke, and fire. Thompson, however, denies us this climax. Release never comes as we watch the train slip the tracks and slide gracelessly down a hillside to squash the Baptist Church Sunday picnic, while the imbecile engineer and scabrous hoboes emerge unscathed. These stranded inhabitants of Thompson's high-noon twilight then merely turn, and resume their ceaseless migration to the next whistle-stop.

*Is it an argument?:*

Yes: the point seems to be that Thompson is hard to pin down as to genre, and the reasons are about his characters, atmosphere, style, and the like.

*Deductive:* I think so; seems like it has suppressed premises here. But I'll treat it as deductive, at least as a first try.

*Is the meaning clear?*

Fairly, but I think paraphrasing would be useful for the purposes of analysis. Author uses metaphor, foreign phrases, and figurative language. Since we're interested in finding out, ultimately, whether it is valid, the argument would benefit from a paraphrase that is crisper and more precise.

*Paraphrase*: (NOTE: statements in [ ] are our additions which are suggested by the text; common knowledge.)

Jim Thompson novels are difficult to pin down as to genre. A typical Western has open space [as the prairie or desert]; potential for personal achievement [society in flux; law often absent or unenforceable]; people who decide and make their own destiny [pioneers, prospectors, gamblers, outlaws, lawmen, etc.]. Thompson's Western space is cramped: characters not self-driven—they are subject to forces outside their control [mental or physical problems, imposed will of others]; they are flawed and fragile, though may be cunning and dangerous.

Typical hard-boiled fiction has sometimes bitter, though self-aware, protagonists who have self-imposed codes of conduct that let them maneuver [more or less successfully] in the often ignored criminal or underclass world where many people are dangerous. The hard-boiled world, however, is part of the everyday world, which is inhabited by people who are unaware of or untouched by, the reality that there is an underclass. Thompson's fictive world is all dangerous, its inhabitants aware that this is all there is, and most have limited success in negotiating this world. They don't have codes of conduct but rules of survival.

Typical suspense fiction [stimulates the reader to anticipation, anxiety, fear or excitement; there is usually a highpoint, followed by resolution]. Thompson's fiction often depresses the reader; there's no anticipation, because readers often think they know what's coming; the approaching ending bears one down, rather than picking one up; the "highpoint" often increases the depression, and there is no resolution—the characters left standing can be expected to continue unchanged.

*Standardize*: (A first approximation; when you work on argument structure, you may find you'd want to revise a bit.)

Final Conclusion: Jim Thompson novels are difficult to pin down as to genre.

Premises seem to be conditional statements and negations:

**If** something is a typical Western, **then** it has open space, potential, and decision makers who take command of themselves in their environment.

Thompson's Westerns have **neither** space, **nor** do characters have potential, nor are they decision-makers nor in command of themselves or their surroundings.

**If** something is typical hard-boiled fiction, **then** it has characters with codes of conduct, and who maneuver in society's underclass.

Thompson's hard-boiled fiction **doesn't** have codes of conduct, **and** it **doesn't** maneuver in society's underclass (because there is just one seedy world!).

**If** something is a typical suspense, **then** it stimulates the reader, leads them in anticipation of the highpoint, then eases them down with a resolution.

Thompson's suspense **depresses** the reader with the sense of inevitability; there is **no** highpoint, **nor** resolution.

*Is Argument Structure Clear?*

From what we did in the standardizing, the structure looks pretty straightforward, but still needs some filling in of intermediate conclusions.

**If** [Thompson's Westerns were typical], **then** it has open space, potential, and decision makers who take command of themselves in their environment.
Thompson's Westerns have **neither** space, **nor** do characters have potential, nor are they decision-makers nor in command of themselves or their surroundings.
[Thomson's westerns are not typical.]

**If** [Thompson's hard-boiled fiction were typical], **then** it has characters with codes of conduct, and who maneuver in society's underclass.
Thompson's hard-boiled fiction **doesn't** have codes of conduct, **and** it **doesn't** maneuver in society's underclass (because there is just one seedy world!).
[Thompson's hard-boiled fiction isn't typical.]

**If** Thompson wrote typical suspense, **then** it would stimulate the reader, lead them in anticipation of the highpoint, and ease them down with a resolution.
Thompson's suspense **depresses** the reader with the sense of inevitability; there is **no** highpoint, **nor** resolution.
[Thompson doesn't write typical suspense.]

[If Thompson novels were easy to pin down as to genre, they would be typical.]
[They are not typical.]

Thompson's novels are not easy to pin down as to genre.

*Is it VALID?*

Argument can be adequately captured argument using LOLA-lite, as it is made up of conditionals and negations.

I won't write out a whole translation scheme here, since the structure is pretty obvious. I'll symbolize in LOLA lite and set up the proof:

1. W → [S & (P & D)]
2. ~[S & (P & D)]
3. H → C
4. ~C
5. T → [E & (A & R)]
6. ~[E & (A & R)]
7. G → [W v (H v T)]      |~G|

This reconstruction yields a very straight-forward proof, using two basic rules and two equivalence rules. Below is the finished proof.

1. W → [S & (P & D)]
2. ~[S & (P & D)]
3. H → C
4. ~C
5. T → [E & (A & R)]
6. ~[E & (A & R)]
7. G → [W v (H v T)]       |~G|
8. ~W                      1, 2 MT
9. ~H                      3, 4 MT
10. ~T                     5, 6 MT
11. ~W & ~H                8, 9 CONJ
12. (~W & ~H) & ~T         10, 11 CONJ
13. ~W & (~H & ~T)         12, AS
14. ~[W v ~(~H & ~T)]      13, DeM/CR
15. ~[W v (H v T)]         14, DeM/CR
16. ~G                     7, 15 MT

Reconstructed in this way, we've shown that we can validly reach the conclusion: Thompson's novels are *not* easy to pin down as to genre.

*Are premises in fact true?*

This question is rarely as straightforward to answer as the question, "Is it valid?" Much depends on the background information you bring to an argument. A bit of skepticism is not a bad thing here—and the *more important the conclusion, the more you want to be sure of the truth of your premises as well as the validity of your argument*. Remember: validity tells us that if those premises are true, the conclusion has to be true as well. So, if we show an argument is valid, *and* we accept the premises as true, we have no other rational choice but to accept the conclusion as true.

For this argument, some research I would want to do if I were unsure of the premises would be:
1. Are these fair characterizations of the Western, Hard-boiled, and Suspense genres? Where could I find out?
2. Is this a fair characterization of Thompson's fiction? How could I find out?
   a. Does it represent his work as a whole?
   b. Is it selectively picking just some part of his work?
   c. Are there any examples from Thompson that would argue against this characterization?
3. (And always) Am I sure that I have fairly represented what the author was trying to say. I should look over the original argument again, and see if I can spot anything missed or misrepresented.

After, you generate such a list of questions, which is about evidence for statements being true:

What kind of evidence?
How much evidence?
Is it relevant evidence?
Is it biased evidence?
Is there any evidence for claims that would argue against the conclusion being true?

This will generally give you clues as to what kinds of reliable sources you'd need to consult to determine if those premises are in fact true.

## ADDITIONAL PRACTICE
Use the Argument-Analysis Tree method to evaluate the following arguments.

*1. Thoughts on Brains*

Our mental states and our brain states are not identical, at least in one sense of the way we use the word 'identical.' We use the word identical when we discover that two things we thought were different turn out to be the same thing. So, we might have thought at one time that the evening star and the morning star were two different things. Later on we discover that they are identical to the planet Venus. Once we know this we would conclude that everything about the morning star would be exactly the same as everything about the evening star; the differences are in name only. This means that everything that happens to one would happen to the other. If this were the case with the mind and the brain, then whenever we had a thought of the number two or scrambled eggs or a car, there would have to be exactly the same brain activity taking place each time. So, if this were the case, then anyone who had a permanent injury to their brain and lost the capacity for these thoughts, would lose those thoughts forever. Since there are cases, however, where people who had brain damage are able to eventually have the same thoughts they had before, these particular thoughts cannot be identical with a particular set of brain states.

*2. Eminent Domain and Private Property*

Despite the lofty claims of legal and political theorists, it is not true that citizens have a right to personal property. Consider a citizen, Bob Johnson, who owns a house at the shore and who pleads with the court that the state has violated his right to property by taking it away from him without his consent. The state argued that they needed to demolish Mr. Johnson's house and one-quarter acre so that they could expand the highway from the shore to the inland in order to ensure the safe evacuation of citizens in case of hurricane or flooding. Mr. Johnson refused to sell, saying that no monetary compensation would be adequate, and that he never wanted to sell. The state invoked the legal doctrine of eminent domain. This doctrine allows for the state to take an individual's property under certain specified conditions, although they must provide just compensation to the individual.

Now, if Mr. Johnson *wanted* to sell, then his right to it still exists. If he *chooses* to take the money the state offered him, then he is exercising his control over it, which is the essence of what it means to own it. To say that Mr. Johnson has a right as a citizen to own something means that, unless he does something while engaged in that right that constitutes him surrendering it or violates the law of the land, he can do what he will with his property. To say that an individual has a right also means that the state is supposed to protect his decision, regardless of what it thinks of him personally or his choices. It is not a right if the state only protects him in those cases when, for its own reasons, it agrees with his decision. To say that Mr. Johnson has a right entails that the state would have to protect this right, even if it doesn't agree with him.

However, the state invoked eminent domain because Bob refused to sell, that is, the state disagreed with his decision. Indeed, by the very nature of such cases, that's the way it's always going to be. In its very essence, the doctrine says that the state only agrees to protect Mr. Johnson's property when they agree with his choices. Therefore, either the doctrine of eminent domain or the concept of a right to personal property can exist but not both. Since, for any number of good reasons, no state will ever deny itself the doctrine of eminent domain, it follows that citizens don't have a right to private property.

## 3. Sacrifice the Bunt!

It's a mistake to hire a manager for any American League baseball team who gives players the okay to bunt. Any manager who does so demonstrates a lack of understanding of statistics and how those stats should guide decisions on the playing field. Every statistical analysis demonstrates that a team's most valuable asset is their at bats. In baseball, as opposed to a sport like football, your defense cannot score for you. You only score when you are at bat. That's why it's essential that you preserve every chance you have to stay at bat. As long as you possess an out, you have a chance to score. Each time you give up an out, you not only radically reduce your opportunities to score, but you increase the odds that the other team will take over at bat. Once they are at bat, you have zero opportunity to score until they lose their three outs.

The idea behind the "sacrifice" is that by giving up an out to move another player along the base paths, the odds will improve that the team will score. While this sounds noble and good, the relevant question is whether the sacrifice is worth the cost. Statistically there are a few situations in which a batter who moves a player from one base to another, but gets out himself in the process, increases the likelihood that the team will score a run. However, obviously better is if a hitter can move the player over *without* also getting out. Hence, using a bunt is meaningful only if the likelihood that the batter will move the player over is *increased* by choosing to bunt. Batters with an average above .075 are more likely to successfully get on base if they swing normally than if they bunt. An unnecessary sacrifice which reduces the opportunities for success is not noble, it is insane.

Any good manager should be aware of these statistics and act accordingly. One thing a manager does have control over is his batting order. Obviously, the only players that should be in the rotation are those who can hit well above .075. If they can't, then they should be pulled as it is not possible for their defensive skills to outweigh their useless offense. The only player that constitutes an exception to this is the pitcher whose value is measured almost entirely in terms of his defense. Of course, this doesn't matter in the American League since it has the DH. Consequently, any American League manager who intentionally tells his hitters to bunt is saying that his own "intuition" or "judgment" is more likely to be correct than a scientific analysis of what will probably take place. While this might be romantic, it's sure not a recipe for success. As for me, I say sacrifice the bunt or sacrifice the managers!

## 4. Segovia's Legacy

Andrés Segovia was a musical legend of the 20th century. To achieve this status requires that the world of music be substantially different from the way it was before they entered the world. The merely great often make important contributions to music in one or two areas. They may be exemplary performers, generate many wonderful compositions, dramatically increase the awareness of their instrument, make advances in technique that set the standard for all who follow, or actively foster the next generation via their pedagogy. A legend does all these things, and thereby changes the musical world forever.

Before Segovia, the guitar was generally viewed as a parlor instrument or one for the local tavern. Although there might be isolated individuals, usually in Spain or South America, who were respected as guitarists, the instrument itself was not thought of as a serious competitor for the violin or piano. Segovia changed people's awareness of the instrument by touring the world. Forgoing small venues for large theaters of up to a 1000 people and embracing the process of recording, he radically increased the world's awareness of the possibilities inherent in the instrument. This created the conditions today in which there are thousands of solo performances and hundreds of guitar festivals every year in almost every country in the world.

He did not, however, simply increase the instrument's popularity. He altered the way the instrument was played and understood by the performers. Differing from predecessors such as Francisco Tarregga, Segovia employed a combination of fingernail and flesh, thereby allowing for greater diversity in tonal colors. Most, if not all, of the most well respected classical guitarists of today employ some version of his approach to the instrument. Through his master classes, his pedagogical studies, and his lectures, he shaped the way guitarists would address, analyze, and perform a piece of music. Many of the most influential classical guitarists of today were students of Segovia. They in turn passed on his lessons through the methods they have created, the instruction books they have written, and the curricula they have established at music schools throughout the world.

Finally, Segovia increased the repertoire of the instrument several fold. He capitalized on the fame and respect he'd garnered to improve what was largely acknowledged to be a relatively thin catalog of substantive musical pieces for classical guitar. He himself transcribed many pieces from the baroque, classical, and romantic masters, including the Bach Chaconne. Although he was not a prolific writer, he actively sought out composers from Mexico, Brazil, England, Italy, and Spain. Because of their respect for him and what he had accomplished they would go on to create a wide range of classical guitar pieces, from small studies to sonatas to orchestral concerti. The music created during this time period now forms the core of the contemporary classical guitarists' repertoire.

Had Segovia contributed in only one or two areas, it is doubtful that the guitar would occupy its current status as one of the most popular instruments in the world. His impact on all these fronts forever altered the way we see the place of this formerly humble instrument in the world of music.

## 5. Don't Let 'em Eat Cake

It would be in the economic best interests of colleges to subsidize the cafeteria stations that serve fruits, vegetables, and salads rather than simply allow market forces to dictate. While consumers make their food selection choices on the basis of brand familiarity, habit, and taste preference, studies indicate that a significant motivational factor in food choice is price. At some price level, one that can be experimentally identified, the incentives of a low cost healthy meal become powerful enough to outweigh other factors. Given that fruits, vegetables, and salads are often relatively expensive to produce and ship to campuses (that is, compared to the unhealthy food choices offered), it would not be possible for the cost of such a meal to reach this optimal price level without institutional support.

The rationale usually given against this view is that subsidizing such venues is economically inefficient. Rather, market forces will invariably result in the most efficient allocation of resources on both the part of the student and the institution. Food chains will engage in competitive bidding with each other for the opportunity to set up shop in valuable campus locations, thereby providing the college with an important revenue stream. Once on campus the contest among the various vendors will be settled by the consumers. All the venues should then be able to draw in enough patrons to maintain their profit margins, either through low prices, or the attractiveness of its offerings, or a combination of both or they will withdraw from the college and leave open an opportunity for another vendor to try its hand. Allowing the market to dictate the food choices on campus will ensure that this form of revenue stream to the college will be optimized.

What is more important to a college's economic well-being than the limited revenue streams provided by food vendors, however, is the level of productivity at the institution. The population of a college that would patronize the on-campus eating venues can be roughly divided up into two camps: employees of the institution (faculty, staff, administrators) and non-employees, i.e., the students. Every statistical analysis available shows that unhealthy eating choices reduce employee productivity by as much as 66%. This is either because of increased absenteeism from sick days or from "presenteeism," the concept that employees might be present but not performing to their optimal productivity level.

Although students are not employees of the college, their level of productivity is also essential to the economic well-being of the college. The greater the student productivity, the better their GPA's and class attendance. The greater the student productivity, the greater the retention rates of the college. The greater the student productivity, the greater their performance on the Graduate Record Exams, and other external measures of student success in learning. Each of these features increases the college's academic standing which thereby increases its competiveness with regard to potential future applicants. Furthermore, as a college's reputation increases, the earning potential of its graduates increases and thereby the potential level of donations by alumni.

It is implausible to suggest that the size of the subsidy would be anywhere close to the figure that would be saved from increased productivity on the part of both employee and non-employees. Hence, it would be economically beneficial for colleges to identify the amount which would be sufficient to generate a significant increase in healthy dietary choices, and set aside part of its operating budget for this purpose.

# Index

Addition (ADD) argument pattern, 257, 269, 273
"all ... except" statements, 99
ampersand, 63, 72, 121, 191
antecedent of a pronoun, 39
antecedent of the conditional, 70–71, 132
Argument-Analysis Trees, 333–34
argument patterns, 253
    Addition (ADD), 257, 269, 273
    Conjunction (CONJ), 256, 269, 271
    Dilemma (DI), 258, 269, 272
    Disjunctive Syllogism (DS), 257–58, 269, 277
    Hypothetical Syllogism (HS), 259, 269, 272–73
    Modus Ponens (MP), 254, 263, 269–73, 276
    Modus Tollens (MT), 255, 269, 272–73
    Simplification (SIMP), 256, 269, 271
argument to absurdity proof. *See* Reductio Ad Absurdum
arguments, 17, 246, 249. *See also* valid argument
    deductive, 13, 245, 248
    definition, 245
    distinguishing from non-arguments, 245
    Horizontal method for displaying an argument, 247, 268
arrow, 70, 72, 121, 191
artificial languages, 13–14, 19. *See also* LOLA; LOLA-Lite
Association (AS), 277
atomic statements, 122–23
atomic statements / compound statements distinction, 121
attributes, 20–21, 25, 33–34

Backward from the Conclusion (tip), 289
Biconditional Breakdown (BCB), 279
biconditionals, 87–88, 121. *See also* conditionals
    using truth tables with, 131
    using truth trees with, 161–63
bivalent logic system, 126
bound variables, 34, 39
"boxed" conclusion, 269
branching, 156, 158, 267
    closed branch, 189
    open vs. closed branches and trees, 166–70
Build a Conjunction (tip), 288

claims or statements. *See* statements
Commutation (COM), 276–77
comparative forms (of adjectives), 99
comparative or relational statements, 25
complex statements
    truth tables and, 135–44
compound statements, 121, 123
    truth-value of, 122
Conclusion in the Premises (tip), 288
conclusion indicators, 246–47, 307
conditional proofs, 282
    provisional assumptions, 283–84
conditionals, 69–73, 121. *See also* biconditionals
    necessary condition, 70
    with quantifiers, 71
    using truth tables with, 132–34
    using truth trees with, 166
conditions
    sufficient condition, 70
Conjunction (CONJ) argument pattern, 256, 269, 271
conjunctions, 63, 99, 121

345

using truth tables with, 128–29
using truth trees with, 154–55
connectives, 128–35
    defined in terms of their truth functions, 126
    operators called, 121
consequent of the conditional, 70–71, 132
consistency, 173–74, 245, 275
consistent statements, 174, 191
contingency, 187–89
contingent statements, 150
contradiction, 187
contradictories, 196
contradictoriness, 187
contradictory statements, 175–77, 191
Contraposition (CONT), 278
contraries, 196
contrary statements, 177–78, 191
CP for Conditional Conclusions (tip), 289
creating an instance, 192–93, 195
Critical Thinking courses, 13, 147, 245

declarative sentences, 13, 15
deductive arguments, 13, 245, 248
"definite description," 97
DeMorgan CRANK (DeM/CR), 280–82
Dilemma (DI) argument pattern, 258, 269, 272
Disjunction/Addition (tip), 289
disjunctions, 75–78, 121
    using truth tables with, 130
    using truth trees with, 158–59
Disjunctive Syllogism (DS) argument pattern, 257–58, 269, 277
Do Something! (tip), 274, 288
double arrow, 88, 191
Double Negation (DN), 277–78
double negations, 162
Duplication (DUP), 280

EE before UC (tip), 305
English language, 7–8, 12–17, 19–21
equivalence, 183–85, 187, 197, 200
equivalence rules, 275, 310
    Association (AS), 277
    Biconditional Breakdown (BCB), 279
    Commutation (COM), 276–77
    Contraposition (CONT), 278
    DeMorgan CRANK (DeM/CR), 280–82
    Double Negation (DN), 277–78
    Duplication (DUP), 280
    Material Implication (MI), 279

equivalent statements, 183
"exactly" statements, 96
exclusive sense of "or," 75–77, 136
Existential Creation (EC), 295–97
Existential Elimination (EE), 298–99
existential instance, 195
existential qualifiers, 58
existential quantifiers, 25, 34–36, 57
    rules for breaking down in a truth tree, 192–93
existential statements, 59, 93, 191–92

Horizontal method for displaying an argument, 247, 268
Hypothetical Syllogism (HS) argument pattern, 259, 269, 272–73

identity, 90, 93, 197
    negated identity relationships, 198
    two-place predicate relation, 91
Identity Introduction (II), 303
Identity Replacement (IR), 304
Identity Rules, 303–04
identity statements, 93, 100, 197, 199
*if and only if*, 87
"if ... then," 69
imperatives, 15
implication, 180–83, 187, 197–98, 245, 248–49
    truth tables to check for, 261
implication rules, 269, 275, 277, 309. *See also* names for individual implication rules
inclusive sense of "or," 75
individual, 20, 23
individual constant, 20
interrogatives, 15

logic
    definition, 11–12
logical falsehoods, 148–49, 187
Logical Language. *See* LOLA
logical truths, 90, 148, 187–90
LOLA, 19–20, 22, 188, 197, 245–47, 261
LOLA-Lite, 121–23, 127, 135, 261
    truth tables and, 154

main operator, 58, 75–76, 88, 135
Material Implication (MI), 279
"mechanical" thinking
    truth tables as example of, 267
method of proof, 267–69
    proof rules for quantifiers, 291–306

Mini-Goals (tip), 288
Modus Ponens (MP), 254, 263, 269–73, 276
Modus Tollens (MT), 255, 269, 272–73
multi-place relations, 29

name, 20
natural language, 12, 14. *See also* English language
    argument patterns in, 253
natural language/artificial language differences, 19
necessary condition, 70
Negated Conclusions (tip), 289
negated existential statements, 191
negated identity relationships, 198
negated quantified statements
    rules for breaking down in a truth tree, 194
negated universal statements, 191
negating the conclusion (tip), 305
negation, 51–53, 55, 57, 59
    using truth tables with, 128
"neither nor," 76, 78
non-identity, statements of, 91
non-reflexivity, 82
non-symmetry, 82
non-transitivity, 82

one-place predicates, 23
"only if," 71
"only" statements, 98
operators, 57, 59, 69
    ampersand, 63, 72, 121, 191
    arrow, 70, 72, 121, 191
    biconditional, 87–88, 121, 131, 161–63
    conditionals, 69, 73, 121, 132–34, 166
    conjunctions, 63, 99, 121, 128–29
    disjunctions, 75–78, 121
    double arrow, 121
    existential qualifiers, 58
    existential quantifiers, 57
    identity relation, 90–91
    negation, 51–53, 55, 57–59
    scope of, 57–58
    tilde, 51–53, 55, 58–59, 121–23
    universal quantifiers, 42, 57, 71, 91, 192
    wedge, 75, 121, 191
"out in the world," 267

parentheses, 65–66, 71
"parenthesis poor," 94, 96
"parenthesis rich," 94

predicate, 20–21, 25–26, 29
    one-place predicates, 23
    order of variables in two-place, 26
premise indicators, 246, 307
proof
    Strategy Tips on how to proceed through, 270, 274, 288–89, 305
proof rules, 309
proof rules for quantifiers, 291–306
    Existential Creation (EC), 295–97
    Existential Elimination (EE), 298–99
    Identity Rules, 303–04
    Quantifier Negation (QN), 301–02
    Universal Creation (UC), 299–301
    Universal Elimination (UE), 291–94
proofs, 269
    conditional proofs, 282–84
proposition, 15–16, 19–20, 22
provisional assumptions
    for conditional proofs, 283–84
    for Reductio Ad Absurdum, 286

quantified statements, 91
Quantifier Negation (QN), 301–02
quantifiers, 33, 191
    more than one quantifier, 45
    with tilde, 55
    universal, 42, 44, 57, 71, 91

range of discourse (RD), 37–39, 46, 59, 72
Reductio Ad Absurdum, 286–87
reflexivity, 79–81, 83, 91
relational predicates, 25
relationships between statements, 249
    consistency, 173–74
    contraries and contradictories, 175–78
    equivalence, 183–85
    implication, 180–83

self-contradictions. *See* logical falsehoods
sentences, 13–15
short-form table technique, 260–61, 263
Simplification (SIMP) argument pattern, 256, 269, 271
statements, 13–15
    "all ... except" statements, 99
    comparative or relational statements, 25
    complex statements, 135–44
    compound statements, 121–23
    equivalent statements, 183
    "exactly" statements, 96

existential statements, 59, 93, 191–92
identity statements, 93, 100, 197, 199
negated existential statements, 191
negated quantified statements, 194
negated universal statements, 191
"only" statements, 98
quantified statements, 91
universal statements, 59, 72, 93, 191–93
statements and declarative sentences
　distinction, 16
Strategy Tips on how to proceed through a
　proof, 270, 274, 288–89, 305
sufficient condition, 70
superlatives, 99
symbolic language, 7
symbolic logic, 12–15
symmetry, 79–81, 83, 91
Symmetry (Sym), 304

tautologies. *See* logical truths
things and their attributes
　distinctions between, 20
three-place predicates, 29
tilde, 51–53, 58–59, 121–23, 191
　and quantifier together, 55
transivity, 80–81, 83, 91
translation, 16–17, 19, 21–23. *See also* LOLA
　alternate translations, 59–61
　LOLA-Lite method, 121–23
truth functional connectives, 122
truth functions, 126
*truth-preserving* rules, 269
truth tables, 123, 126–27, 147, 260–62
　"build up," 154
　complex statements and, 135–44
　example of "mechanical" thinking, 267
　rule governing existential statements in, 199
　using to demonstrate a connective's function, 128–34
truth trees, 153–72, 187–88, 191, 249–50
　branching, 156, 158, 267
　"break down," 154

closed branch, 189
negated identity statements and, 198
open vs. closed branches and trees, 166–70
rule for breaking down existential qualifiers in, 192–93
rule for breaking down negated quantified statements, 194
rule governing identity statements, 199–200
rules for breaking down universal statements in, 193
truth-value of a compound statement, 122
truth-value of a proposition, 126
truth values, 263
two-place predicates, 25–26
　order of variables in, 26

Universal Creation (UC), 299–301
Universal Elimination (UE), 291–94
universal quantifiers, 42, 44, 57, 71, 91, 192
universal statements, 59, 72, 93, 191–92
　rules for breaking down (in a truth tree), 193
"unless," 75

valid argument, 17, 249–50, 269
valid argument pattern, 254
validity, 248, 261, 308
　property of an argument, 249
　seeing, 268
validity and short-form tables, 260–61
variables, 21, 26
Vertical method
　displaying an argument by, 249, 251, 268

wedge, 75, 121, 191
well-formed formula. *See* WFFs
WFFs, 22–23
What kind of statement is the conclusion? (tip), 305
what's in the box stays in the box, 284, 287

# from the publisher

A name never says it all, but the word "broadview" expresses a good deal of the philosophy behind our company. We are open to a broad range of academic approaches and political viewpoints. We pay attention to the broad impact book publishing and book printing has in the wider world; we began using recycled stock more than a decade ago, and for some years now we have used 100% recycled paper for most titles. As a Canadian-based company we naturally publish a number of titles with a Canadian emphasis, but our publishing program overall is internationally oriented and broad-ranging. Our individual titles often appeal to a broad readership too; many are of interest as much to general readers as to academics and students.

Founded in 1985, Broadview remains a fully independent company owned by its shareholders—not an imprint or subsidiary of a larger multinational.

---

If you would like to find out more about Broadview and about the books we publish, please visit us at **www.broadviewpress.com**. And if you'd like to place an order through the site, we'd like to show our appreciation by extending a special discount to you: by entering the code below you will receive a 20% discount on purchases made through the Broadview website.

Discount code: **broadview20%**

*Thank you for choosing Broadview.*

Please note: this offer applies only to sales of bound books within the United States or Canada.

The interior of this book is printed on 100% recycled paper.